D0537036

TARGET EARTH

TARGET EARTH

DUNCAN STEEL

Reader's Digest

THE READER'S DIGEST ASSOCIATION, INC.
Pleasantville, New York/Montreal

A READER'S DIGEST BOOK

Conceived, designed, and produced by
Quarto Publishing plc
The Old Brewery
6 Blundell Street
London N7 9BH

SENIOR PROJECT EDITOR   Nicolette Linton
ART EDITOR   James Lawrence
PICTURE RESEARCH   Duncan Steel, Laurent Boubounelle
PHOTOGRAPHER   Paul Forrester
ILLUSTRATORS   David Asher, David Crawford, Don Davis, James Garry, William K. Hartmann,
Kuo Kang Chen, Scott Manley, Michelle Stamp
INDEXER   Dorothy Frame

ART DIRECTOR   Moira Clinch
PUBLISHER   Piers Spence

Reader's Digest Project Staff
PROJECT EDITOR   Kimberly Ruderman
SENIOR DESIGNER   Carol Nehring
ASSOCIATE EDITORIAL DIRECTOR   Marianne Wait
SENIOR DESIGN DIRECTOR   Elizabeth L. Tunnicliffe
EDITORIAL MANAGER   Christine R. Guido

Reader's Digest Illustrated Reference Books
EDITOR-IN-CHIEF   Christopher Cavanaugh
ART DIRECTOR   Joan Mazzeo
DIRECTOR, TRADE PUBLISHING Christopher T. Reggio
EDITORIAL DIRECTOR, TRADE   Susan Randol

First published in 2000 by
The Reader's Digest Association, Inc.
Pleasantville, New York 10570-7000

Text copyright © 2000 Duncan Steel
Concept, design, and layout copyright © 2000 Quarto Publishing plc

All rights reserved. Unauthorized reproduction, in any manner, is prohibited.

Library of Congress Cataloging in Publication Data

Steel, Duncan, 1955 –
    Target earth: the search for rogue asteroids
    and doomsday comets that threaten our
    planet / by Duncan Steel; with an afterword by Sir Arthur C. Clarke.
    p. cm.
    ISBN 0-7621-0298-5
    1. Asteroids—Collisions with Earth. 2. Comets—Collisions with Earth. I. Title.

QB651 .S76 2000
523'.44—dc21   00-041503

Reader's Digest and the Pegasus logo are registered trademarks of
The Reader's Digest Association, Inc.

Printed in China by Leefung-Asco Ltd.

QUAR.DOCO

Contents

Foreword

Target Earth, the title of this book, describes our planet in its cosmic context. If one phenomenon has dominated the sculpting of the many planets and moons we have investigated, forming the vast craters that pockmark their surfaces, it is impacts by asteroids and comets.

Such hugely energetic events are not only important for geology. They are also of significance for biology. We know only of life on Earth, but it is thought possible that microbial life could be transferred between the planets – from Earth to Mars, say, or vice-versa – within rocks blasted into space by massive impacts. Certainly many meteorites are proven to have originated on the Moon, and some on Mars. Whether the latter contain signs of primordial Martian life is another matter.

Leaving aside the transfer of bacteria from planet to planet, we now know that asteroid and comet impacts over the eons have influenced the evolution of terrestrial life. Although there is still much scientific debate on this matter, it is widely believed that the extinction of the dinosaurs (along with the majority of other living things) 65 million years ago was triggered by such an impact. The residual scar, a crater around 120 miles (200 kilometers) wide, has been identified on the Yucatán Peninsula in Mexico. Debris from the explosion has been found spread all over the world.

This is not the only impact to be linked to a mass extinction event. The southern end of Chesapeake Bay seems to have been shaped by an impact 35 million years ago. Just a few months ago it was announced that another large crater had been recognized in Australia, one that may be linked to the biggest extinction of all some 250 million years back.

These were all truly cataclysmic events, releasing energy equivalent to hundreds of millions of megatons of TNT. Although these super-impacts release phenomenal energies, they occur very infrequently, at intervals of tens of millions of years. A more pressing danger is represented by those smaller impacts that release less energy, but occur more frequently.

About every 100,000 years – we cannot be sure exactly how often, but that is a reasonable approximation – the Earth will be hit by an asteroid a half-mile or a kilometer in size. This would kill a very large number of people. The many variables, such as the impact speed, the object's composition, whether it hits land or ocean, and so on, mean that the precise consequences cannot be forecast. But certainly we would expect hundreds of millions, perhaps billions, of people to die. The statistical calculations are quite simple. They show that there is a greater chance of an individual (you, reading this book) dying through an asteroid striking our planet than in a jetliner crash. Indeed asteroid impacts dominate all other natural hazards (such as earthquakes, hurricanes, floods,

Saturn

Amor

Sun
Earth

Aten

Mars

Apollo

Comet
Halley

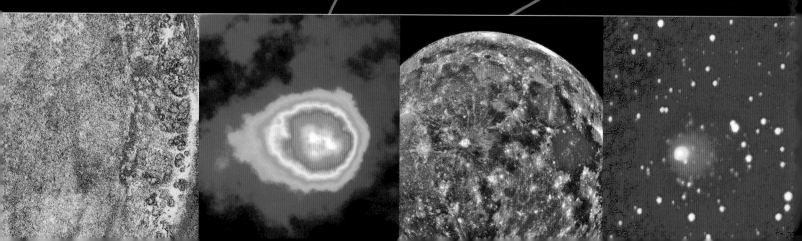

Jupiter

Earth

Comet Hale-Bopp

century, but our realization of the awful consequences demands that we pay the premium to ensure our safety. Having recognized the level of danger, not to carry out the necessary search program would be foolhardy and irresponsible.

In *Target Earth* our colleague Duncan Steel argues this case at length. Using a wide variety of images showing craters on the Earth and elsewhere, photographs of asteroids and comets, and other results from the latest scientific investigations, he presents the evidence that our planet is indeed a target in a celestial shooting gallery.

If humans are to continue to thrive on Earth, and perhaps eventually expand into space, we must avoid being hit by some piece of cosmic detritus that could put an end to civilization. We have the scientific and technological ability to do this. We hope you will be persuaded that we *should*.

tornadoes, volcanic eruptions and so on) in terms of your chance of meeting an early death.

This is all based on statistics and probabilities, and on our present state of ignorance. Over the past decade or so we have learned a great deal about asteroids and comets, allowing us to estimate how often, on average, they slam into Target Earth. But we don't know *when* the next impact is due. It might not be for millennia; but it could be tomorrow. It is the timing of the next impact catastrophe of which we are ignorant. But this ignorance can be remedied.

Cosmic impacts differ from the other natural hazards in this respect. We cannot predict the vagaries of the weather that result in storms. But we *are* able to foresee an asteroid or comet impact years ahead of the event. If we knew a large rock from space was due to hit our planet, we could do something about it. The first step would be to find it, and that is the aim of the Spaceguard Foundation.

We are talking about a cosmic insurance policy. As with automobile insurance, we pay the premium but hope that we will never need to make a claim. Most likely there is *not* an asteroid or comet due to hit our planet within the next

Dr. Andrea Carusi
(President, The Spaceguard Foundation);
Institute of Space Astrophysics,
Rome, Italy

Dr. Brian Marsden
(Vice President, The Spaceguard Foundation);
Director, Minor Planet Center,
Harvard-Smithsonian Center for Astrophysics,
Cambridge, Massachusetts, U.S.A.

August 2000

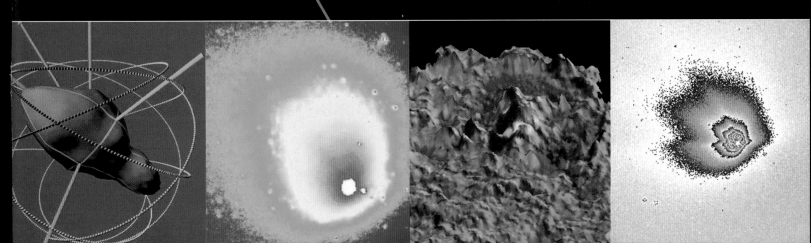

The Cosmic Beehive

We live on a planet enveloped in a swarm of cosmic projectiles: asteroids and comets. These lumps of rock, metal, and ice pose a substantial and surprising hazard to the well-being of each and every one of us.

This is not in the realm of science fiction. This is science fact. Researchers have gained a better – and more frightening – comprehension of the cosmic impact danger in recent years. In this book we explore the hazard these celestial missiles represent, and what could and should be done to ensure that mankind does not go the way of the dinosaurs.

This makes for scary reading in places. We live not on Planet Earth, but on Target Earth.

Hearing the Buzz

Earth's inhabitants have watched comets move sedately across the night sky since the dawn of time. Vivid apparitions occur perhaps once in a decade, interspersed with Comet Halley's appearances every seventy-six years. Although seen only once a century or so, truly amazing comets made their mark upon ancient peoples, and all the great civilizations have left us breathless accounts of their sightings. These were phenomenal spectacles, with tails stretching halfway across the sky, and were often bright enough to be seen during broad daylight.

The appearance of a great comet would have had an influence on our ancestors. In the popular mind they were regarded, like eclipses, as a sign of changing fortune. A wily leader might rally his armies by declaring a comet to be a good omen. More often, soldiers would interpret such a celestial visitor as a sign of evil foreboding. Just how evil comets can be was not recognized for many centuries, until scholars such as Edmond Halley showed that they could slam into the Earth – with potentially calamitous consequences.

But there were things that Halley did not know about

three centuries ago. Comets are obvious in the sky because of the clouds of vapor that surround them as sunlight evaporates some of their ice. This makes them bright. There are also bare rocks in space, and those are far dimmer and much harder to see. We call them asteroids.

The first asteroid was discovered on the first day of the nineteenth century. In 1898 astronomers found the first such object capable of coming close to our planet. In the 1930s three asteroids were recognized whose trajectories implied that they could possibly hit Earth. In the following decades a handful more were added. But in the past few years the number of known Earth-approaching asteroids has escalated rapidly, and we have at last woken up to the threat they pose to us. We do indeed live in a cosmic beehive.

Feeling the Sting

Looking out into space is astronomy. But point your telescope at the Moon and you can see the connection between astronomy and geology. The pockmarked lunar surface tells us that comets and asteroids have shaped the face of the Moon. Why should the Earth be any different? Every so often our planet must be struck a hammer blow. When that happens we'll all feel the sting.

The Moon and Earth are different. Our companion is inert, and so it retains the craters that are the evidence of impacts from millions and billions of years ago. The Earth, though, is active. Our atmosphere produces weathering, which quickly erodes craters. Other processes such as earthquakes and volcanoes also scrub the evidence. But geologists have learned ways to identify the vestigial scars left by previous strikes, and hundreds of terrestrial impact sites are known.

Cosmophobia

This presents us with a problem: is the Earth a safe planet? Or is it possible that the sky can fall on our heads? Perhaps there is need for a little cosmophobia – a realization that all is not benign out there in space. The following pages reveal why.

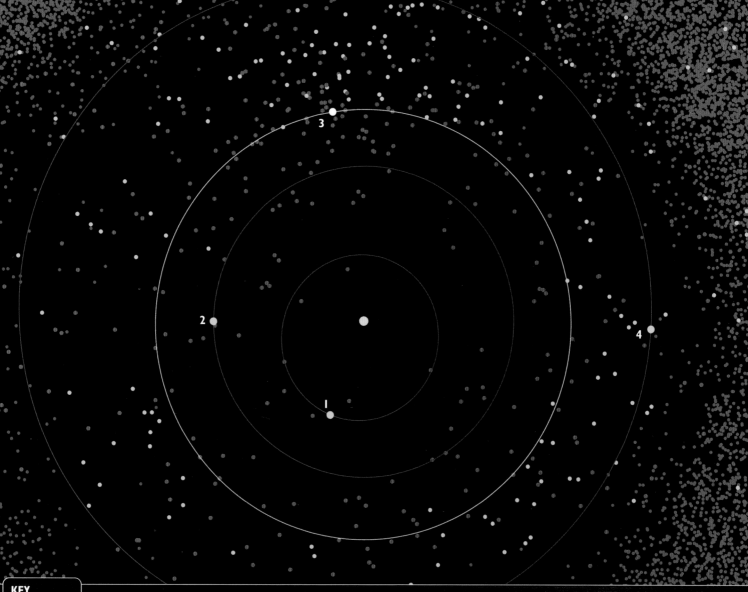

Above Maps of the solar system usually show only the Sun and the planets. Astronomers now know there is a swarm of asteroids sharing space with us. Each dot indicates a known asteroid, but we are sure that there are many more awaiting discovery. The red dots represent asteroids that cross the Earth's orbit, making impacts feasible.

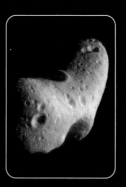

Far left Comets are easy to find due to their accompanying clouds of gas and dust that reflect a lot of sunlight.

Left Eros, the first near-Earth asteroid to be discovered, as photographed by NASA's NEAR-Shoemaker satellite.

Extending the Solar System

Five planets may be easily observed with the naked eye: Mercury, Venus, Mars, Jupiter, and Saturn.

In fact, the word *planet* is derived from the Greek for wanderer, because the planets move relative to the fixed background of stars. Ancient civilizations such as the Babylonians also considered the Sun and Moon to be planets, because they too appear to be in motion. That makes seven in all, from which our seven-day week is derived, with each day named for one of the planetary gods.

These ancients thought the solar system was complete – comets were simply occasional visitors – and this belief persisted for millennia. This is nicely illustrated by the fact that when Uranus was discovered in the late eighteenth century, it was initially assumed to be a comet.

Uranus Spotted

William Herschel, an organist and composer from Germany, moved to the city of Bath in England to take up a musical position. But his private hobby was astronomy.

While scanning the sky with his telescope in 1781, Herschel spotted a bright and fuzzy object. It appeared to be moving slowly across the sky, relative to the stars, implying that it was a member of the solar system. Herschel thought it must be a comet, and so it was proclaimed to be one.

But comets brighten as they come closer to the Sun, and develop tails. Herschel's discovery did not: seen through the telescope, it remained a small disk. Measurements of its shifting position over months indicated that it was on a near-circular orbit, not the elongated path expected of a comet. Eventually the truth dawned: it was an unknown planet.

In recognition of his discovery, Herschel received not only a knighthood but also a liberal monetary allowance from George III, the king of England. In return, Herschel and the British astronomical community called the planet George's Star. But European astronomers would have none of this, insisting on naming the planet after the mythological god Uranus. And so it has remained to this day.

Neptune Predicted

Had the full extent of the solar system been discovered? Many thought so, but Uranus refused to behave itself. Precise measures of its position over the following fifty years showed slight wobbles, which could not be accounted for by the gravitational tug of the other planets alone.

In the 1840s, two mathematicians independently theorized that an unseen planet must be responsible, and set about calculating its possible location and mass. In France, Urbain Leverrier communicated his solution to various European observatories. In England, John Couch Adams, an undergraduate student at Cambridge,

Above William Herschel, who discovered Uranus in 1781.
Below Copernicus drew the planets in orbits going around the Sun, the outermost circle representing the sphere of the fixed stars. No planets beyond Saturn were known in his era, five centuries ago.

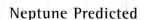

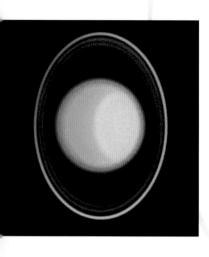

Left Uranus and its rings photographed with the Hubble Space Telescope showing the correct orientation; Uranus has its polar axis tipped over, and the rings orbit above its equator.

tried to convey his own results to the Astronomer Royal, Sir George Biddell Airy, who did little with them. James Challis, professor of astronomy at Cambridge University, began a slow search using Adams's prediction, and actually saw the new planet without realizing it. On the European mainland, progress was much quicker. Neptune was discovered from the Berlin Observatory in September 1846, much to the chagrin of many British astronomers who saw this as a blow to a nation regarding itself as preeminent in the field.

Discovery of Pluto a Fluke

But the story does not end there. As time passed it was found that neither Uranus nor Neptune followed their computed paths precisely, indicating the possibility of a ninth planet farther out. There was much written about this idea in the early 1900s, and at the Lowell Observatory in Arizona a photographic search project began. As a consequence, Clyde Tombaugh discovered Pluto in 1930.

But it was a fluke. Pluto was near the predicted position, but obviously too tiny to explain the wobbles of Uranus and Neptune. A planet with a mass at least six times that of Earth had to be responsible, and Pluto is smaller even than our Moon. This hatched the idea of Planet X: a notional tenth planet orbiting beyond Pluto. For years astronomers searched in vain for this hypothetical body. But the real reason for the wobbles did not emerge until 1992.

Estimates of the masses of Uranus and Neptune came from observations of their many moons. The estimates were derived from the orbital periods of the moons and their distances from the parent planets. But from distant Earth such measures were of limited accuracy. Far superior determinations were feasible from radio tracking data for the Voyager 2 spacecraft, which

passed Uranus and Neptune in 1986 and 1989 respectively. The deviations of the satellite's path gave an accurate indication of the planets' gravitational pulls, revealing that previous mass values were out by a few parts in a thousand, one too high, the other too low. With the new measures the wobbles were eliminated, and the observed paths of the planets over the past couple of centuries agree with the newly calculated values.

What does this mean? It implies that Pluto was found in a deliberate search that was based on a false premise. It also means that there is no Planet X – at least not having a mass as large as the Earth – within 10 billion miles (16 billion kilometers) of us.

Above A vista of Neptune returned by the Voyager 2 spacecraft in 1989. We still do not know whether this vast ball of gas has any solid surface. Wispy white clouds are seen circulating the planet, largely colored blue by the methane in its atmosphere.

Nocturnal Apparitions

Few who have seen a magnificent comet illuminating the night sky ever forget the sight.

The great tails accompanying a comet give the objects their title. The word comet is derived from the Greek for hairy star.

The vast majority of comets are faint, requiring a telescope to be seen, but some are far brighter. The most famous of all, Comet Halley, has been sighted on its return every seventy-six years since the third century B.C. Earlier records most likely exist, perhaps on Babylonian clay tablets, waiting to be uncovered one day. Astronomers have good reasons for thinking that Comet Halley has followed a path much like its present trajectory for at least ten millennia.

Above Comet Halley appeared in 1066, as depicted in the Bayeux Tapestry recording the Battle of Hastings in England.

Comet Swift-Tuttle was also recorded over 2,000 years ago. We have special reason to be interested in this comet, because its orbit passes so close to the Earth that every year we fly through tiny pieces of debris, called meteoroids, left in its wake. This produces the Perseid meteor shower, a free fireworks show occurring between August 12 and 14.

But could the comet itself – a lump of dirty ice about 10 miles (16 kilometers) across – slam into our planet? The answer is yes, although it will not do so soon. In 2126 it will miss Earth by two weeks, and will not come closer during the third millennium. We do not need to be alarmed about Comet Swift-Tuttle, but we should take the Perseid meteor shower as a sign that comets may be dangerous. Little bits of them are hitting us continuously.

Ancient Superstitions

All civilizations have witnessed great comets. The ancient Chinese recorded them as "broom stars" as a way of describing their appearance. But with that notion comes the idea that comets are harbingers of change – sweeping out the old to make way for the new.

The most famous comet in literature was seen in 44 B.C. and is associated with the assassination of Julius Caesar. In Shakespeare's play, Calpurnia urges her husband not to go to the Senate on the Ides of March, telling him:

When beggars die, there are no comets seen;
The heavens themselves blaze forth the
death of princes.

Above It has often been suggested that the Star of Bethlehem was really a comet, as depicted in this medieval painting by Giotto.

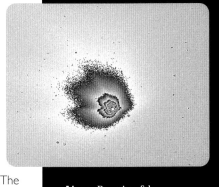

Above By using false color coding, astronomers can trace out detail within the bright coma surrounding a cometary nucleus. This was Comet Hale-Bopp in 1996.

Left In a more conventional photograph, Comet Hyakutake stretches across the night sky above city lights.

Below A false color rendition can also delineate the structure within the multiple tails of a comet.

It is not quite true that a comet indicated Caesar's doom. The records show that a bright comet did appear that year, but not until months later. The association may have been made because the comet blazed across the sky in the summer, around the time of the special games held to mark Julius Caesar's death.

But this does not mean that comets were invariably thought of as omens of disaster. Comets, like eclipses, might be interpreted as signs of good as well as evil, depending upon the superstitions of the culture and the beliefs of the period. Unrest and discontent might lead to a particular comet being viewed fatalistically. But if the people's morale were high – perhaps after an excellent harvest, for example, or the birth of a royal son – then the same celestial object would have been interpreted favorably.

Such beliefs continue today to some extent. For example, in some wine-growing regions it is claimed that a prominent comet will lead to a fine vintage and the bottles are labeled "Comet Wine." Of course, the producers themselves might not share that belief, but if the public seizes on the story, then it affords the shrewd marketer an opportunity to obtain a premium price for the season's wines.

Great Comets

In history, "naked-eye comets" have arrived at a typical rate of one every two or three years. A century or two ago – before artificial lights lit up the dark night sky and restricted the number of visible stars – comets were much more familiar to the majority of people. They would note a faint comet for a few weeks, and then it would be gone.

But every so often a great comet would appear. It might be intrinsically large and active, so that it appeared bright even though distant from the Earth. Or it might be an object happening to pass relatively nearby, giving it a long tail and bright head. Two comets from the seventeenth century, those of 1664 and 1680, are well remembered. The nineteenth century was blessed with phenomenal comets in 1843, 1860, 1880, 1882, and 1887. They all deserved the description "great," although there were many other spectacular visitors.

Comparatively speaking, comet sightings during the early twentieth century were sparse. There were several prominent comets, such as Kohoutek in 1973 and West in 1976, but the comet-watcher's world was set alight in the last few years when the Hyakutake and Hale-Bopp comets appeared. The latter was one of the largest comets to arrive in centuries, with a nucleus about 25 miles (40 kilometers) in size. If it had passed close to the Earth it would have been even more spectacular. But, as this book shows, we would certainly prefer big comets and asteroids to keep their distance.

Left sequence The most photographed comet in history, Hale-Bopp is shown here moving against spectacular foreground scenery.

Truth and Consequences

In medieval Europe it was widely believed that comets brought disease and pestilence to the land. But out of this age of superstition also came the first inklings that comets could do far worse.

The seventeenth century saw great advances in astronomical knowledge. The Copernican theory, which declared that the Sun was the center of the solar system since all other bodies have it as the focus of their orbits, gained widespread acceptance. Johann Kepler published his laws describing planetary movements. In the latter half of the century, Isaac Newton unveiled his laws of

Right Edmond Halley was the first to predict the return of a comet – the famous object that now bears his name.

gravitation and motion, which made it possible to calculate the orbits of comets.

At that time, the belief among astronomers was that comets were vagabonds from interstellar space, passing the Sun on the one and only occasion we would have a chance to view them before they flew off into space never to return. But Edmond Halley computed the orbits of several comets and showed them to be gravitationally bound to the Sun. Nowadays, comets are named after their discoverers, but Halley did not "discover" the comet named after him: we have observations of it stretching back to 240 B.C. It bears his name because he recognized that comets seen in 1531, 1607, and 1682 were one and the same object, since the orbits he computed for them were virtually identical. On that basis, he predicted that the comet would return, long after his death, in 1758. Comet Halley has appeared on schedule ever since.

Cometary Collisions

Edmond Halley's calculations rendered the orbits of many comets that had been precisely recorded in preceding years. But those orbital parameters had implications beyond the fact that

KEY
BLUE AND WHITE LINES — orbits of planets: all move counterclockwise/prograde
GREEN LINES — orbits of comets that move counterclockwise/prograde
RED LINES — orbits of comets that move clockwise/retrograde

3 Earth	D Schwassmann-Wachmann 3
5 Jupiter	
6 Saturn	E Hartley 2
A Swift-Tuttle	F Encke
B Halley	G Wild 2
C Tempel-Tuttle	H Wirtanen
	J Crommelin

Left The orbits of some of the known periodic comets, several of which are targets of spacecraft in the next few years.

some comets are periodic, returning again and again after set intervals.

One parameter is called the perihelion distance, which is the closest approach a comet makes to the Sun. The Earth, having a near-circular orbit, maintains more or less the same distance from the sun — about 93 million miles (150 million kilometers) — and we call this distance one Astronomical Unit (AU). Comets, though, have more egg-shaped paths, and this elongation is described by astronomers as the "eccentricity" of an orbit.

An eccentricity of zero describes a circle, while an eccentricity of one implies a parabola. This is an orbit that is on the limit of being gravitationally tied to the Sun. If the eccentricity is larger than one, the orbit is unbound, and an object on such a trajectory would fly off into interstellar space, never to return.

Periodic comets such as Halley that return repeatedly have eccentricities between zero and one. The orbital periods, though, do not depend on the eccentricity but on the size of the orbit. We know of many comets that return every five to twenty years. The shortest period belongs to Comet Encke, which comes back every forty months.

Divine Intervention

Because of their elongated or egg-shaped orbits, some comets pass closer to the Sun than the Earth (they have perihelion distances of less than 1 AU), despite the fact that they spend most of their lifetimes far away in the outer reaches of the solar system. Halley realized that this meant a collision with Earth was a possibility, and he presented this idea to the Royal Society of London in 1694. His particular concern was the origin of the biblical deluge and "the great agitation that such a shock must necessarily occasion in the sea." He suggested that the Caspian Sea and various other huge lakes had been produced by cometary collisions, and his theories were later taken up by others.

Almost immediately, Halley's ideas got him into trouble with the established Church. As a result, his discussions were not published for another thirty years, by which time he was Astronomer Royal. Having the common people believe that comets set plagues upon them as manifestations of divine displeasure was a valuable weapon for the Church hierarchy. The need to appeal to divine intervention for protection from cataclysmic impacts was not so useful.

Above A medieval depiction of the biblical flood. Halley and others of his era insinuated that a cometary impact may have given rise to the original legend. **Background and below** An allegorical nineteenth-century French representation of the comet threat to the Earth.

The First Asteroids

When eighteenth-century astronomers found a new planet in the outer solar system, they were not entirely surprised: it had been anticipated for some time that there should also be one between Mars and Jupiter.

Above Johann Bode, whose law of planetary distances led astronomers to expect a planet between Mars and Jupiter.
Below Giuseppe Piazzi, the discoverer of the first asteroid.

This was on the basis of simple numerology – the significance of which is still debated by researchers investigating the origin of the solar system. The numerology appears in what is usually called Bode's Law, named for the German astronomer Johann Bode, although it was his compatriot Johann Titius who devised it in 1766.

They had noted that the planetary distances from the Sun could be described by a simple arithmetical expression, with the exception of a gap between Mars and Jupiter, which the calculations said should contain a planet.

How Big Is a Planet?

On the night of January 1, 1801, in Sicily, Giuseppe Piazzi spotted a new planet orbiting between Mars and Jupiter – just where Bode's Law said it should be. This discovery – now named Ceres – was greeted with acclaim. But it was soon realized that Ceres was far too small to be classified as a proper planet, since it is only about 567 miles (913 kilometers) across, just a quarter the diameter of our Moon.

Further confusion ensued in 1802 when a second, slightly smaller body was found between Mars and Jupiter: it was given the moniker Pallas. In 1804 and 1805, the first two bodies were joined by Juno and Vesta. There was a hiatus until 1845, when Astraea was found. Thereafter, using the bigger and better telescopes becoming available, along with the onset of photography, astronomers started to find inhabitants of this previously "empty" band at an ever-increasing rate. By the end of the nineteenth century, hundreds of bodies were known. Today, there are over 15,000 in the sequence that begins with

1 Ceres, 2 Pallas, 3 Juno, and 4 Vesta; millions more have been observed on single nights and may eventually be added to the master list.

The Main Belt

The International Astronomical Union insists that these bodies be referred to as minor planets, although it is more common to call them asteroids. The latter term derives from their appearance through a telescope. They look like stars (*aster* means star), rather than the nebulous, fuzzy profile presented by a comet.

The vast majority of asteroids orbit the Sun in what is called the main belt, a wide band of stability between Mars and Jupiter. The majority remain locked in orbit, but every so often one breaks free of the belt, allowing it to approach the Earth. These are the rogue asteroids, the ones we need to worry about.

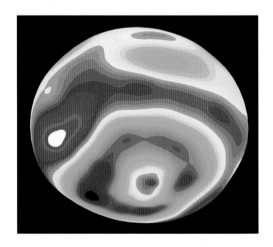

Above The shape of asteroid Vesta captured by the Hubble Space Telescope. Vesta is about 323 miles (520 kilometers) across, but not quite spherical.

Right This map of the inner solar system shows all known asteroids in the positions they occupied at the start of the year 2000. The main belt is the dense cluster of green dots between Mars and Jupiter. Red and yellow dots depict asteroids with rogue orbits making impacts possible on the terrestrial planets (Mars, Earth, Venus and Mercury). It's the red ones that are a danger to us.

KEY
1 Mercury
2 Venus
3 Earth
4 Mars
5 Jupiter

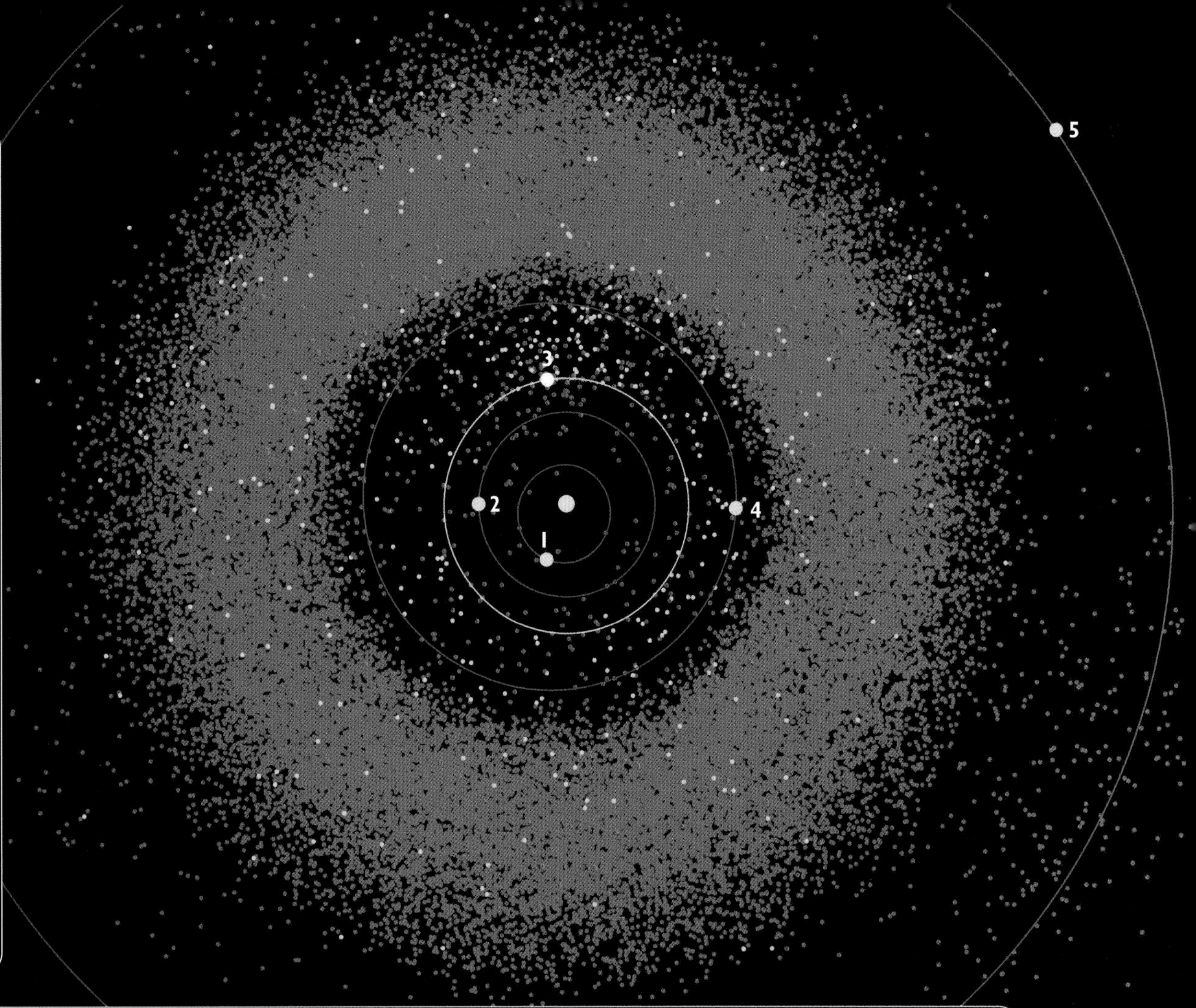

Bode's Law of Planetary Distances

Bode's Law comes in various forms. The following is the most common.

Start with 0, followed by 3, and then a succession of doublings of the previous number:

0	3	6	12	24	48	96	192	384	768

Next add 4 to each number:

4	7	10	16	28	52	100	196	388	772

Divide each number by ten:

0.4	0.7	1.0	1.6	2.8	5.2	10.0	19.6	38.8	77.2

... and compare to the actual average solar distances of the planets, in AU:

Mercury	Venus	Earth	Mars	?	Jupiter	Saturn	Uranus	Neptune	Pluto
0.39	0.72	1.0	1.52	-	5.20	9.55	19.2	30.11	39.5

When the law was seriously discussed in the 1770s, only the planets as far as Saturn were known, and the "gap" at 2.8 AU seemed suspicious. When Uranus was found in 1781 close to the "predicted" position, many were convinced that the law represented reality, and assumed that another planet lay waiting to be discovered between Mars and Jupiter. The discovery of Ceres in 1801 seemed to fit the bill perfectly. But the truth is much more mundane, as indicated by the later discoveries of Neptune and Pluto, whose orbits do not fit the predictions. Bode's Law has no physical basis: it is pure coincidence. The universe is full of simple relationships that happen to fit this or that series of numbers, with no hidden meaning in explaining its laws.

Too Close for Comfort

No sooner had the first asteroids been discovered in the main belt than some astronomers reached the conclusion that similar objects might pass close to the Earth. Some might even hit it.

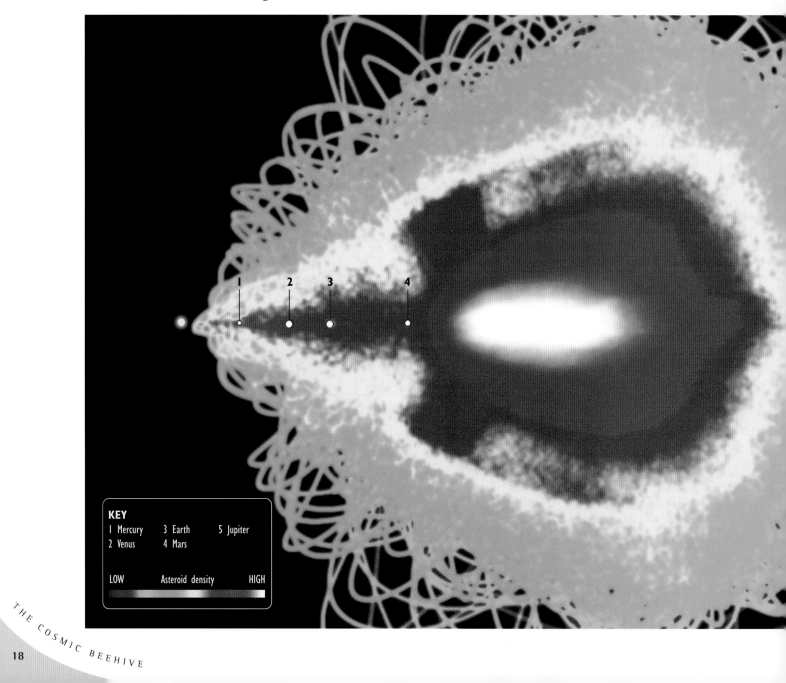

KEY

1 Mercury   3 Earth   5 Jupiter
2 Venus   4 Mars

LOW   Asteroid density   HIGH

Below left This cross-section through the solar system shows the density of orbits of known asteroids. The white dots indicate the position of the planets moving outwards from the Sun, far left. The main belt, shaded white and violet, has the highest asteroid density. The Earth lies within the next band, shaded red.

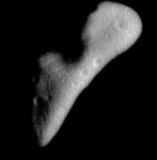

Left and right Through a telescope, asteroids are just pinpricks of light. Recent spacecraft visits have shown us the shapes of some asteroids such as Eros (left) and Ida (right).

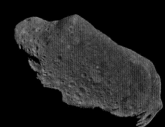

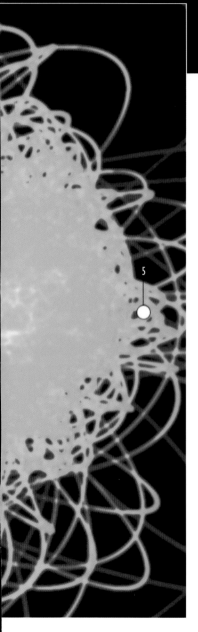

In the last years of the eighteenth century, Halley's ideas about the potential of comets to cause impact catastrophes had resurfaced, notably through the writings of the great French scientist Simon Laplace. (Laplace is also credited with inventing the concept of a Black Hole: a celestial object of such phenomenal mass that light cannot escape from it.) France was the place for such ideas, in the tumult and aftermath of the revolution. Baron Georges Cuvier, one of the founders of paleontology, suggested that there had been several gross extinctions of life on Earth, with few species surviving. The biblical flood was perhaps the latest of them.

In this climate it was obvious to point to cometary impacts as the cause of the extinctions, but there was a problem. Prevailing opinion indicated that the Earth was only a few thousand years old, based on biblical chronology. Some scientists, such as Count Buffon – another Frenchman, who in the middle of the eighteenth century had reasoned that a massive comet plunging into the Sun led to the formation of the planets from the rebounding blob of material – believed that the Earth might be hundreds of thousands of years old. But all estimates were below a million years. It was quite simple to calculate that a random comet had about a one in a billion chance of striking the Earth on each orbit. So given the observed number of comets, it seemed unlikely that more than one comet

might have hit the planet since its formation. The notion that there might be large, dark rocks in space that could do the job – whether you call them asteroids or minor planets – provided a solution to the puzzle, if you believed that the extinctions postulated by Cuvier and others had been caused by impact catastrophes.

The First Near-Earth Asteroids

But none of the first asteroids known to astronomers had paths coming anywhere near the Earth: they were all safely tucked in between Mars and Jupiter and traveling in orbits that were not expected to alter substantially over time scales of millions of years. All that changed in 1873 when the first asteroid crossing the orbit of Mars was identified. Twenty-five years later Eros was found. Eros was the first of the near-Earth asteroids – and it was a massive one.

Eros cannot hit the Earth because its orbit does not cross our path, but its discovery brought about renewed speculation on the possibility that asteroids might pose a danger to our planet. (In fact, calculations published in 1994 by a team from the Nice Observatory in southern France have shown that the orbit of Eros might evolve in a way that makes a terrestrial impact possible within the next million years.) Then, in 1932, a small asteroid named Amor was spotted. It has a path that brings it close – uncomfortably close – to Earth.

Earth-Crossing Asteroids

Before astronomers had a chance to absorb the implications of Amor, they were unsettled by the discovery of another asteroid just weeks later.

Now called Apollo, it was the first asteroid known to have an Earth-crossing orbit, making an impact feasible, at least in theory. Soon it was joined by Adonis: it became clear that there were other threatening objects out there.

Some argued that this pair had trajectories that kept them far from the Earth itself, but that argument was undermined in 1937 when another asteroid, Hermes, was observed for just a few days as it whizzed by at a distance not much farther from the Earth than the Moon. Hermes, which is about a half mile (800 meters) across, disappeared into the cosmic haystack and was lost because the data collected on it were insufficient to plot a definitive orbit about the Sun. We simply do not know where it is,

although we are sure that it comes back to pass us by every so often. This means it could hit the Earth at virtually any time, without warning.

We've Got a Problem

Between the 1940s and 1960s several more Earth-crossing asteroids were discovered, essentially by accident. They were picked up on photographic plates shot through wide-angle telescopes for other purposes. Astronomers interested in studying distant stars and galaxies took to calling asteroids the Vermin of the Skies, because they "spoiled" otherwise unblemished celestial photographs.

But at the same time other scientists were becoming convinced that the craters of the Moon had an impact origin, and that these structures also dotted the surface of the Earth. It didn't take much to calculate the energy involved in excavating a relatively small crater, such as Meteor Crater in Arizona, and to compare the answer to the yields of the most powerful atomic bombs that had been tested. Knowing how little of the sky had been scanned, astronomers could estimate how many Earth-crossing asteroids were out there, waiting for us to find them before they found us. That led to calculations of how often our planet might be struck. The answer was uncomfortable, in view of the havoc that would be caused by a one-mile (1.6-kilometer) projectile releasing energy equivalent to a million megatons of TNT when it hit the ground at 20 miles (32 kilometers) per second. All of a sudden the message was clear: we have a problem.

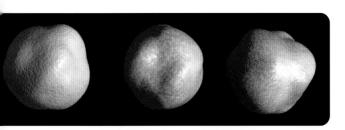

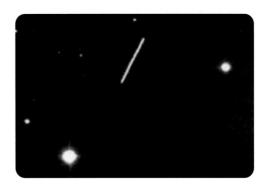

Below left The shape of the small asteroid 1998 KY26, which flew close by the Earth in that year, is clearly shown in these radar profiles.

Bottom left Earth-approaching asteroids leave telltale streaks on long-exposure photographs, and this has often led to their discovery. As the Earth spins, the telescope is rotated at the correct rate to compensate, and so the images of stars appear as static dots. But during an exposure lasting an hour, an asteroid will have moved tens of thousands of miles, and betrays its presence by drawing a line across the image.

Right This map of the orbits of near-Earth asteroids known at the start of the year 2000 superimposed on the planetary orbits shows the degree of congestion in our solar system. Red lines show asteroid orbits that cross that of the Earth. Yellow lines show asteroid orbits coming closer to us than Mars, but unable to strike our planet at present. The fact that a few yellow lines appear to cross our orbit is a perspective effect produced by mapping their tilted orbits onto the Earth's orbital plane.

KEY	
1 Mercury	3 Earth
2 Venus	4 Mars

THE COSMIC BEEHIVE

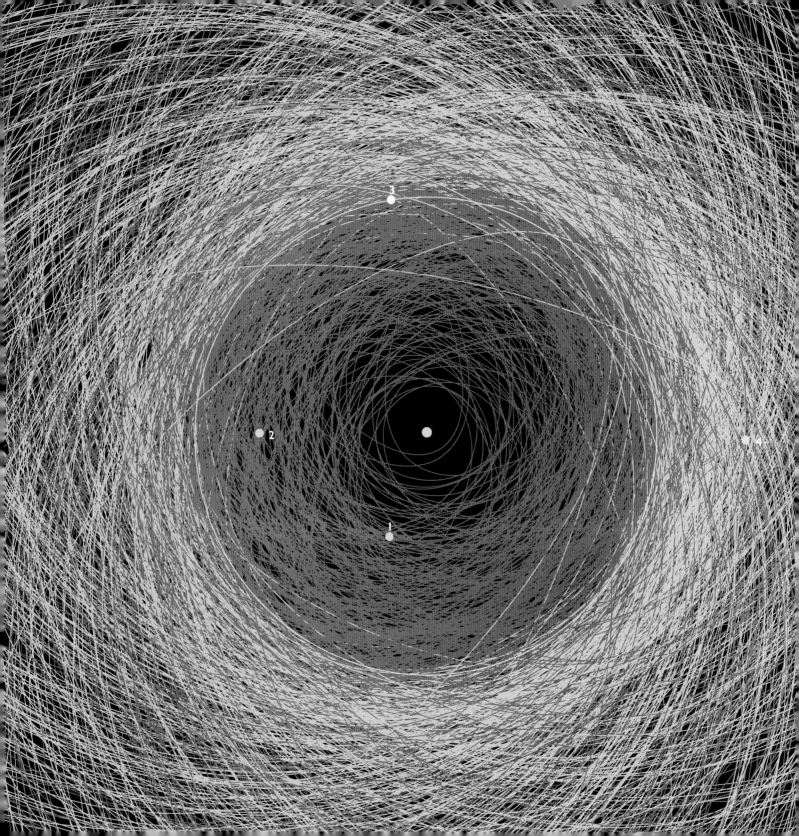

Comets All Over

Most comets have highly elongated orbits, taking thousands of years to return – if they ever do.

Asteroids, on the other hand, have orbits that are fairly small. They typically take a few years to complete a loop around the Sun, with some notable exceptions.

Although a handful of comets have orbital periods as short as five years, most are much longer. Even Comet Halley, at seventy-six years, counts as short-period by the cometary norm. The majority of comets recorded have paths so elongated that they take thousands or even millions of years to complete an orbit. Such objects tend to be seen once only, blazing in the sky for a month or two before disappearing again into the depths of space. A good example is Comet Hale-Bopp. Although we have no record of its previous appearance, we know it was about 4,000 years ago that it last visited the inner solar system. During its apparition in the 1990s it was decelerated slightly by the gravitational tugs of the planets. This reduced its orbit size, so that it will be only a couple of thousand years before it comes back. But that's still a long, long time.

The orbits of the planets are prograde, that is, they all circle the sun in a counterclockwise direction. Asteroids do likewise – although with a broader spread of orbital tilts to the planetary plane, as seen on page 18. To look for asteroids, point your telescope close to that plane, because that is where most of the potential targets lie.

Comets are not constrained in space in this way. They travel in all directions, with every possible orientation to the plane of the Earth's orbit about the Sun. In principle, you can look for new comets in any part of the sky.

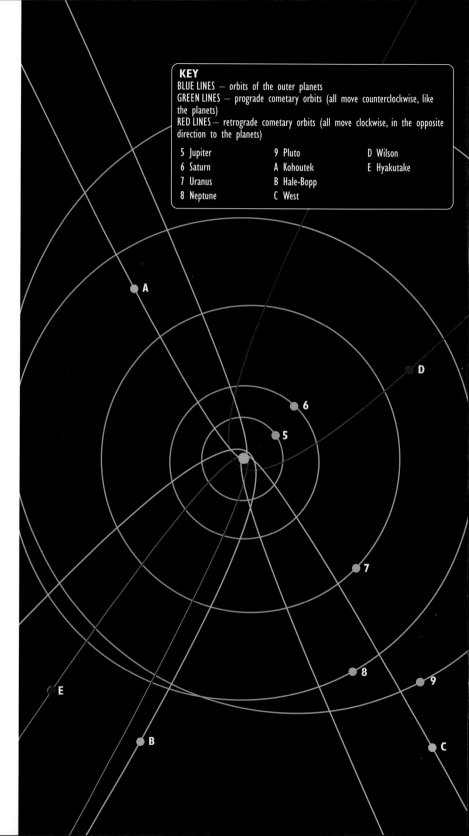

KEY
BLUE LINES — orbits of the outer planets
GREEN LINES — prograde cometary orbits (all move counterclockwise, like the planets)
RED LINES — retrograde cometary orbits (all move clockwise, in the opposite direction to the planets)

5 Jupiter	9 Pluto	D Wilson
6 Saturn	A Kohoutek	E Hyakutake
7 Uranus	B Hale-Bopp	
8 Neptune	C West	

The Oort Cloud

The realization that comets can arrive from any direction is an important clue to their origin: it tells us that their source must be spherical, surrounding the solar system.

Another clue comes from the fact that most comets are close to the edge of being gravitationally bound to the Sun. That is why their orbital periods are so long: they have sufficient speed and energy to travel far from the Sun before its gravity brings them to a halt and drags them back again. That is, they are members of our solar system, but until they were thrown into orbits allowing them to approach

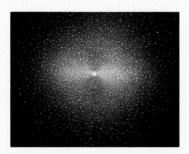

the Sun and brighten – enabling us to see them – they must have spent eons slowly circling in space way out beyond Neptune and Pluto.

This reservoir of comets, disturbed every so often in a way that drops comets in toward us, is usually called the Oort cloud. It derives its name from the Dutch astronomer Jan Oort, who proposed its existence in 1950, although similar ideas had been expressed earlier by Ernst Öpik. The cloud is huge, containing many billions of comets that typically orbit at a distance of 10,000 to 100,000 AU from us, a quarter of the range of the nearest stars.

Opposite Most comets observed have greatly elongated orbits that take thousands or even millions of years to complete. Shown here are some of the most famous long-period comets seen in recent decades. Also illustrated are the near-circular orbits of the outer planets (Jupiter, Saturn, Uranus, Neptune, and Pluto) and the smaller orbits of the inner planets (Mars, Earth, Venus, and Mercury).

The Galactic Connection

Comets were formed with the rest of the solar system about 4.5 billion years ago. We think that they accumulated from the debris left over when the giant outer planets were produced. Soon after, the gravitational slingshot action of Saturn, Uranus, and Neptune flung them farther out into space, where they formed the Oort cloud. Since then, the huge number of comets in this vast reservoir – about a billion millions (1,000,000,000,000,000) of them larger than a mile (1.6 kilometers) in size, according to one estimate – have mostly continued to plod around the frigid depths of space without ever coming close to the planets.

But every so often they get a kick. For example, a passing star may barge through the cloud, knocking many of them into new trajectories that make them fall in toward the planets. Or one of the massive clouds of gas and dust that inhabit the galaxy might gravitationally unsettle some of the comets as the solar system wanders by. Another effect is caused by the Milky Way as a whole: we know that the Sun oscillates up and down through the galactic plane about every 30 million years, and that this causes a bending and compression of the Oort cloud,

which in turn drops a few comets our way.

On a human scale, though, this all takes time. A comet dislodged from the Oort cloud will take a couple of million years to plummet toward us. In fact, there may well be one or more comets already on a path that will end with an Earth impact within a few million years. But comets so far away are impossible to detect.

Changing Orbits

Many comets pass us by repeatedly, giving them multiple chances to score a bull's-eye. Comet Halley is the obvious example, but there are dozens of comets that return again and again.

How do these comets get into their periodic orbits? Comets coming from very large orbits may be diverted into smaller ones when they pass close to a planet, especially Jupiter. Passing in front of the planet will slow the comet down, injecting it into an orbit of shorter period. But if it passes behind the planet, the comet will be accelerated into a larger orbit, and ejected into interstellar space. The giant planets both enhance and reduce the impact hazard. They trap some comets into smaller orbits, giving them multiple impact opportunities. Others are thrown out, never to return.

Trans-Neptunian Objects

The Oort cloud is not the only source of comets – a vast new belt of interplanetary debris has been discovered much closer to home.

Above Even using the Hubble Space Telescope we can see little detail of distant Pluto, pictured here with its moon Charon.

The discovery of the Oort cloud began with astronomers' observations of comets. They deduced the comets' source through an analysis of their orbits. There is no chance of seeing comets in the Oort cloud directly, because at a distance of a light-year the individual objects are far, far too dim to be detectable. It is difficult enough to pick up a one-mile (1.6-kilometer) lump of bare rock and ice with the world's largest telescopes when it is as close as Saturn, and impossible to see a comet over a thousand times more distant.

Later theoretical work suggested that these comets were formed within the region of the outer planets, and then thrown out into the chilly depths. But what about comets that might have been produced from debris just beyond Neptune? With no massive bodies out there capable of slinging them farther, the comets should remain in place, plodding around the periphery.

Beyond the Blue Planet

The idea that there should be a belt of comets or protoplanets stretching some distance beyond Neptune was proposed by Dutch-American astronomer Gerard Kuiper in 1951 – although historical digging has shown that an Irish astronomer, Kenneth Edgeworth, introduced the concept a couple of years earlier.

Although the Oort cloud is too far away to be detected directly, there was hope that this new and much nearer belt might be within the realm of our telescopes. It could be claimed that Pluto, about 1,400 miles (2,300 kilometers) across, is simply the largest and brightest member of the group. If there were others at

least 100 miles (160 kilometers) in size, then they should be detectable from Earth.

Confirmation of this hypothesis came in late 1992, when the first trans-Neptunian object (TNO) was discovered by a team led by David Jewitt of the University of Hawaii. Since then some hundreds have been found, ranging up to 300 miles (500 kilometers) or so in size, inhabiting what is now called the Edgeworth-Kuiper belt, in honor of its proposers.

Are these comets, or are they minor planets (asteroids)? As a matter of record, the International Astronomical Union officially designates them as minor planets. Minor planets are given numbers as well as names, such as 1 Ceres, 433 Eros, 1862 Apollo, 2101 Adonis, and so on. In the late 1990s the number 10,000 was approaching, and professional astronomers proposed that while Pluto should keep its name, it could be added to the list of minor planets as 10000 Pluto. This would have recognized the fact that it is not really big enough to rank with the eight major planets. But this proposal upset some members of the public, who felt that Pluto was being downgraded, so the International Astronomical Union backtracked and left things the way they were.

Instability Rising

But are all these TNOs really minor planets? The point is that we expect them to be made not of rock and metal and other inert constituents, but largely of ice, solid carbon dioxide, methane, ammonia, and other volatile chemicals, which are known to comprise much of the mass of comets. That is, the TNOs are really giant comets. But at the extreme

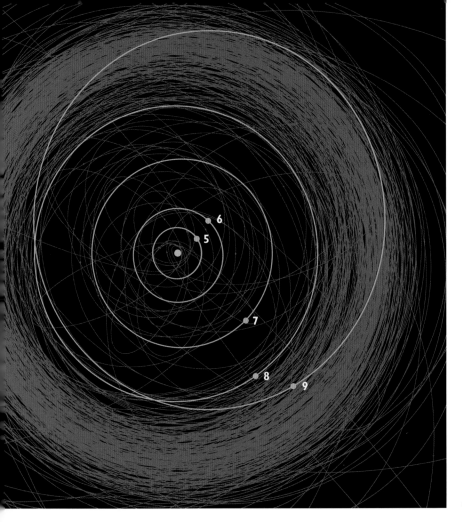

Many of the TNOs happen to be on stable orbits. Pluto has an orbit that crosses Neptune's, and yet they cannot collide because two Pluto orbits take precisely as long as three Neptunian orbits, and they keep their distance. This is an example of a resonance. Similarly, many TNOs have resonant orbits, protecting them from any marked deviations. But others do not.

It seems inevitable that from time to time a massive TNO must enter the planetary region, possibly to be flung inward toward the terrestrial planets. Comets are somewhat fragile and often break apart, leaving several large fragments. In principle a 100-mile (160-kilometer) TNO could fall sunward and disintegrate into a million separate pieces each 1 mile (1.6 kilometers) across. Then we'd have chaos.

Above A minor planet out beyond Neptune gives away its presence by moving during the hour or so between images of deep space taken with a large telescope. The other bright spots are stationary stars and galaxies.

Above A map of all the outer solar system minor planets known at the beginning of the year 2000. Most of these are safely beyond Neptune, but there are also several on paths that bring them in to cross the planetary orbits. Called Centaurs, these are discussed on page 27.

these are discussed on page 27.

KEY

5 Jupiter	8 Neptune
6 Saturn	9 Pluto
7 Uranus	

---

temperatures so far from the Sun – at least a couple of hundred degrees below zero – their icy contents remain frozen.

If such an object fell sunward, heat from the Sun's radiation would start to evaporate that ice, and we would have the most phenomenal comet imaginable. Comet Hale-Bopp was around 25 miles (40 kilometers) across, but in this case we are talking about a nucleus perhaps ten times that size and a hundred times the surface area. It is the surface area that determines how much of the ice can evaporate or sublime to spawn the vast cloud of vapor (the cometary coma) that reflects sunlight to our eyes and telescopes.

Trojans and Centaurs

● Asteroids and comets whose paths cross the planetary orbits are in most cases unstable.

Over timescales of thousands or millions of years they will approach a planet and either hit it or be diverted into a different orbit. This tells us that the planet-crossing objects we see now cannot have been there since the solar system formed 4.5 billion years ago: they can only be transient occupants of their orbits. Dynamic change is occurring constantly, generally over eras far longer than human lifetimes, but short on an astronomical scale. There are certain exceptions to this rule however. Pluto,

for example, crosses Neptune's orbit, but the planets cannot collide because, as mentioned on page 25, there is a 3:2 resonance between their orbital periods. They move in step, so they never come close to each other. Even simpler would be a 1:1 resonance, and we see many instances of that in the solar system.

Because of its huge gravitational field, Jupiter is the dominant planet, and most objects with trajectories crossing the jovian orbit are highly unstable. There are, though, many asteroids that occupy basically the same orbit as that giant planet. These are called the Trojans.

Trojan asteroids are in a 1:1 resonance with a planet. That is, they have the same orbital period, 11.86 years in the case of Jupiter. This protects them from approaches to the planet so long as they are far enough away. Such detailed celestial mechanics leads to the idea of *Lagrangian* points. Two of these, labeled L4 and L5, are 60 degrees behind and ahead of Jupiter in its orbit. An asteroid or comet occupying those zones will be protected and can remain there indefinitely. Rather than being located

Far left Comets are often seen breaking into a few fragments. This artist's impression shows a cometary nucleus shattering into many small bodies.
Left In 1995, Comet Hale-Bopp released a daughter fragment, as shown in this Hubble Space Telescope image.
Below The best-known broken-up comet of recent times was Shoemaker-Levy 9, as shown in this montage. Its chain of fragments slammed into Jupiter in July 1994.

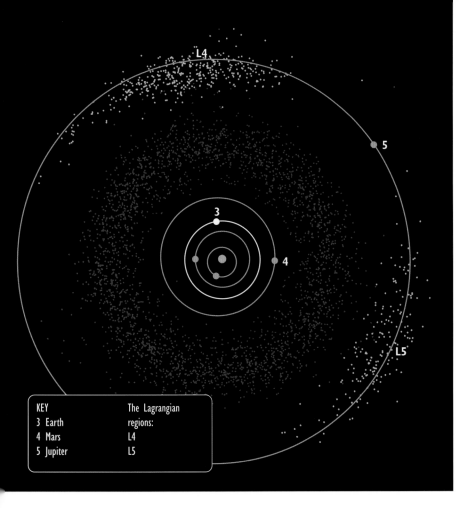

KEY
3 Earth
4 Mars
5 Jupiter

The Lagrangian regions:
L4
L5

Left This map shows the positions of all minor planets larger than 30 miles (50 kilometers) in size at the start of the year 2000. The clouds of Trojan asteroids at the L4 and L5 points ahead of and behind Jupiter are obvious.

specifically at the L4 and L5 points, the many known Trojan asteroids tend to oscillate about the general regions, keeping their distance from Jupiter. Apart from these jovian Trojans, an asteroid was discovered in 1990 which seems to be a Mars Trojan, moving in the same way relative to that planet. More recently we have become aware of an Earth-crossing asteroid moving in a complicated fashion that allows it to keep its distance from us in much the same way.

Among the Planets

Such orbits as those mentioned above are stable, but they are the exceptions. More often, planet-crossing orbits are unstable, even chaotic. They change in major respects with high

frequency. For example, typical Earth-crossing asteroids have trajectories that we can predict with precision for only a century or so. From the perspective of preserving ourselves and the next few generations, that is fine. But to understand the ebbs and flows of the smaller bodies in the solar system, and how they have affected life on Earth, we need to consider longer timescales than this. In this respect, the Centaurs are important. We have previously discussed how large trans-Neptunian objects might enter the outer solar system and be deflected in toward the terrestrial planets, breaking into myriad potentially lethal fragments as they travel. And we do actually see massive asteroids and comets among the outer planets. These are the Centaurs.

Seventy years ago a substantial asteroid, Hidalgo, was found with an orbit crossing Jupiter and Saturn. It was an oddity, but it was largely ignored. Interest in such objects was rekindled in 1977 when Chiron — around 200 miles (320 kilometers) across — was found on a path crossing both Saturn and Uranus. It was numbered as 2060, a minor planet. But in the 1980s, as Chiron approached its perihelion, observations showed signs of outgassing. Some sort of highly volatile material was being evaporated, meaning that Chiron was really a comet.

During the 1990s several more Centaurs were found (see the orbit map on page 25). All were big, all were on unstable orbits, and all were listed as asteroids, although one might presume that they are really icy bodies. It is inevitable that some will eventually fall into the inner solar system and break apart.

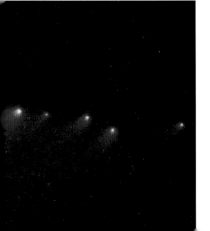

Craters, Craters, Everywhere

In the preceding chapter we considered some of the history of the basic – but terrifying – concept that the Earth is subject to an occasional cosmic impact, which would surely leave a very substantial crater. Then we reviewed some of the evidence gathered in recent years indicating that there are asteroids and comets essentially everywhere in the solar system. Indeed, one could say that the solar system ends where the comets stop – at the edge of the Oort cloud. And we also know of comets falling into the Sun, as we'll see. Nowhere is safe.

If such cosmic projectiles sculpt the faces of the planets, then it follows that there should be craters everywhere. The next step in our tutorial is to take a tour of the solar system, using the plethora of detailed images collected by spacecraft during the past few decades. We will find that, no matter where we look, every picture tells the same story.

Impact Energies

These unguided missiles hit the planets hard and fast: the typical impact velocity of an asteroid hitting the Earth is around 12 miles (19 kilometers) a second. Although it might sound fast, that is actually a relatively low speed: the Earth is moving at between 18 and 19 miles (around 30 kilometers) a second – about 67,000 miles (107,000 kilometers) per hour – in its orbit around the Sun. The impact speed is lower because asteroids, like all the planets, orbit in the same direction around the Sun: counterclockwise when viewed from the north. This is called prograde motion.

Many comets, though, have orbits proceeding in the opposite direction, and are said to be retrograde. Comet Halley is a pertinent example: if it were to run into us, it would collide at 41 miles (66 kilometers) a second. For comets arriving from the Oort cloud, the average impact speed would be about 34 miles (55 kilometers) a second, almost three times as high as the asteroids. Because the impact energy increases as the square of the speed, energies released on impact by comets are typically eight times as high as those released by asteroids. On one hand, this makes comets intrinsically more dangerous. On the other hand, comets are far easier to spot in the night sky, and they are far fewer in number.

Hypervelocity objects made of stone, metal, or ice are inert in themselves, but they have a great deal of explosive potential due to their kinetic energy (energy of motion). Even at a speed of 2 miles (3 kilometers) per second, a lump of rock possesses more kinetic energy than the chemical energy available from the same mass of TNT, and cosmic impacts on Earth occur at far greater speeds than that.

How much energy is released in a major impact? Consider a 1-mile (1.6 kilometer) asteroid with the density of common rock. Imagine that it slams into the Earth at 12 miles (19 kilometers) per second. Simple physics shows that the explosive yield would be around 240,000 megatons of TNT.

How does that compare with our nuclear weapons? Well, it's 3,600 times higher than the most powerful hydrogen bomb ever tested, or almost 20 million times the yield of the atom bomb dropped on Hiroshima in 1945.

How big a hole would it make? As a rule of thumb, craters are 10 to 20 times the size of the impactor, depending on the speed and density of the projectile, the strength of the target rock, and so on. We might expect each 1-mile asteroid hitting a planet to leave a crater around 15 miles (25 kilometers) across. So where are these craters?

Right Nuclear weapons liberate huge amounts of energy, but they are dwarfed by the explosive power released by even a moderately-sized cosmic impact.

Above A comet flies past the Earth-Moon system. The lunar surface is dominated by craters resulting from hypervelocity collisions by comets and asteroids. Why imagine that the Earth, seen here in the distance, is safe from such impacts?

Our Pockmarked Neighbor

If you want to gather evidence for impact craters, you don't need to look far. Just go outside on a clear night with a small telescope or a decent pair of binoculars and look at the Moon.

Everywhere you look there are craters. Even a modest telescope will make over 30,000 lunar craters visible.

Strange Ideas

Despite repeated suggestions that the lunar craters might have impact origins, it was many years before this theory was widely accepted. Many geologists remained unconvinced until rocks were returned from the Moon after the first Apollo landings in 1969. The telltale signs of shock metamorphism in the rock finally presented inarguable evidence that impacts were responsible for the craters.

For decades the popular belief was that they were volcanic in origin. On Earth there are many circular calderas produced by volcanoes, and it was natural to assume that the structures on the Moon, which look broadly similar, must have the same origin. But we now know that the Moon is a dead body with a solid core – all volcanism ceased there eons ago. Most of the craters were formed long after the lunar volcanoes became extinct. In any case, the craters are far bigger than any volcano could produce.

Over the centuries a host of bizarre suggestions have been made to explain the origin of the craters. Perhaps the strangest is that they might be similar to coral atolls. Yet it is easy to see how this notion came about.

The Waterless Seas

Looking at the Man in the Moon (or the Rabbit in the Moon, to some cultures), the features are delineated by visible dark-bluish patches. The ancients thought these were oceans – an error compounded by Galileo when he saw them through his newly invented telescope. As a consequence, they are called the maria (singular, mare), meaning seas. For example, Apollo 11 landed on the Mare Tranquillitatis, or Sea of Tranquility. We now know that the maria are vast impact basins formed soon after the Moon solidified, and subsequently filled with basalt from volcanic eruptions while there was still a liquid interior. That is why

Above The Apollo 11 lander, bringing back the first samples of Moon rock.
Top right Many craters on the moon are old and have been eroded by subsequent impacts. The bright rays of ejecta from this crater, named Giordano Bruno, show that it is relatively young.
Right An artist's impression of the impact that might have formed the Moon: a Mars-sized body slams into the primordial Earth, and some of the rock is ejected into orbit before coalescing to produce our natural satellite.

Below left The pockmarked face of the Moon, from Apollo 17 in orbit.

the maria look dark and blue: they are composed of a different sort of rock from the lunar mountains.

When Charles Darwin went on the voyage of the Beagle to the Pacific in the 1830s, he returned with detailed observations of the coral atolls he had examined in the tropics. These are usually large, circular rings of coral around a small island, and have a similar appearance to many lunar craters. Although astronomers knew by then that the Moon lacked water, it seemed reasonable that the craters might be the remains of coral atolls left high and dry when the Moon lost its water, assuming it once possessed oceans. The theory was wrong, but plausible in the context of the knowledge of the times.

The Moon's Impact Origins

The face of the Moon has been shaped by impinging asteroids and comets, but in addition our natural satellite owes its very existence to a tremendous, cataclysmic impact. It is believed that the terrestrial planets were formed through the gradual accumulation of large lumps of debris – protoplanets, planetesimals, asteroids – that were the remnants of a huge nebula of gas and dust from which the Sun condensed 4.5 billion years ago. It seems the early Earth was struck by another massive object about the size of Mars. This resulted in a vast amount of rock, many times the present lunar mass, being thrown into space. Some would have escaped, but some remained in orbit. Computer simulations have shown how that ejecta could have accumulated into a single lump, which we now call the Moon. This theory for the Moon's origin – the "Giant Impact Hypothesis" – accounts successfully for the major features of our natural satellite, such as the ways in which its composition differs from the Earth. The story of the Moon, then, is dominated by impacts from start to finish.

Lunar Gardening

There are those who find it possible to believe that, while the Moon has certainly been pockmarked by marauding asteroids, somehow – thanks to divine intervention, perhaps – the Earth has avoided being hit.

You might think that the terrestrial atmosphere protects us (the Moon is naked to space) and there is some truth to that. Our atmosphere is surprisingly effective at stopping small projectiles from reaching the ground intact. Rocky asteroids smaller than 100 yards or meters in size, and icy comets below 200 yards or meters across, are mostly shielded out. They blow up at altitudes of several miles when the frictional force imposed on them exceeds their strength. Although this cataclysmic blast can cause damage on the surface of the Earth, it does not result in craters. The Moon, however, has no such protection.

Craters Great and Small

Every year about 40,000 tons of meteoroids and interplanetary dust particles enter our atmosphere. The really tiny dust grains slow down without melting, and then gradually fall to the surface where we can collect them from repositories such as the Antarctic. Run a magnet along the gutters on your house, and the chances are that you'll find several tiny metallic meteorites, washed down from your roof by the rain. You'll need a magnifying glass, though, as they'll be smaller than a hundredth of an inch across. Larger grains ablate, or burn up, in the atmosphere, producing the meteors or shooting stars that most people have seen occasionally.

The Moon, though, has no such shielding atmosphere. So the meteoroids and dust particles punch into its surface unimpeded, still traveling at tens of miles or kilometers a second. Their kinetic energy causes them to evaporate on impact, leaving behind tiny craters and pits. Over eons the Moon's surface has been churned over by these arrivals in a process known as lunar gardening.

While you can see tens of thousands of Moon craters through a telescope, up close many more are revealed. In fact, the closer you look, the more craters you see. From orbit, a good camera system shows that they continue down in size until the resolution of the optical system is reached.

When the Apollo astronauts landed on the surface of the Moon, they found myriad small craters that were not visible from orbit. And when the rocks were brought back to laboratories on Earth and examined with suitable microscopes, there were microcraters all over the surfaces that had been exposed for millennia.

In essence, there is only one form of weathering that occurs on the Moon, imposed by impactors of all dimensions. Once a crater is formed, it stays there until another impact occurs and wipes it out. In this regard, we talk of saturation cratering. The lunar surface is largely saturated with craters, so that each new crater-forming impact wipes out (roughly) one old one.

But what evidence do we have that lunar cratering is occurring now? First, there are various recently formed craters visible from the Earth, such as Tycho and Giordano Bruno. It is obvious that these are comparatively young. For example, they have bright ejecta blankets around them that have not yet been darkened by subsequent impacts from microscopic projectiles, or overlapped by fresh, moderate-sized craters.

Second, the Apollo astronauts left seismometers on the surface, and those sent

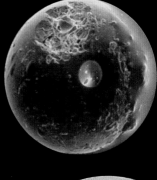

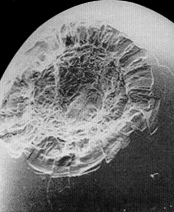

Above Samples of Moon dust. These are spheres of glass produced in large impacts that have themselves been pitted later by tiny grains hitting them. **Right** Apollo 12 landed near the earlier unmanned Surveyor 3 probe, shown here. The lunar surface is clearly covered by craters of all sizes, from large to small.

back data indicating that meteoroids the size of a car regularly slam into the Moon.

Third, amateur astronomers have actually seen impacts occurring as flashes detected while using modest telescopes. Such flashes had been reported for years, but were dismissed as figments of the imagination, or faults in the recording system. During the Leonid meteor storm in November 1999, however, several observers reported such flashes at precisely the time when the Moon was passing through the core of the meteoroid stream. That seems proof positive that the phenomenon is real.

Indirect evidence for these small impacts comes from a tenuous atmosphere of sodium and potassium atoms that astronomers have detected around the Moon. Indeed, during the 1999 Leonid meteor storm a tail of sodium, similar to a comet tail, was seen stretching away from the Moon.

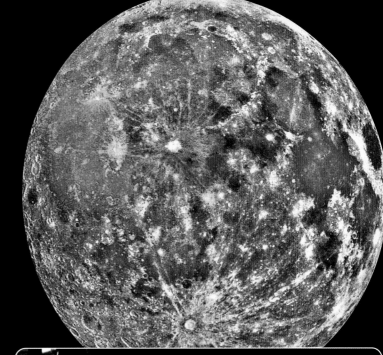

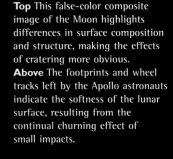

Top This false-color composite image of the Moon highlights differences in surface composition and structure, making the effects of cratering more obvious.
Above The footprints and wheel tracks left by the Apollo astronauts indicate the softness of the lunar surface, resulting from the continual churning effect of small impacts.

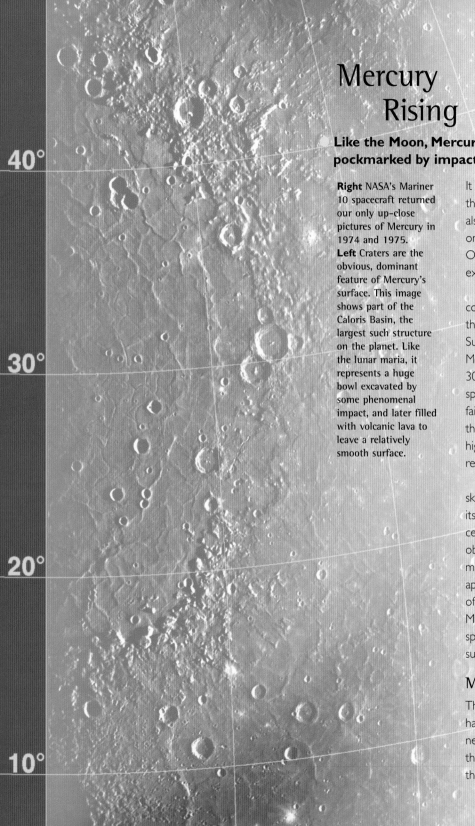

40°

30°

20°

10°

Mercury Rising

● **Like the Moon, Mercury is also pockmarked by impact craters.**

Right NASA's Mariner 10 spacecraft returned our only up-close pictures of Mercury in 1974 and 1975.
Left Craters are the obvious, dominant feature of Mercury's surface. This image shows part of the Caloris Basin, the largest such structure on the planet. Like the lunar maria, it represents a huge bowl excavated by some phenomenal impact, and later filled with volcanic lava to leave a relatively smooth surface.

It is clear that the asteroid flux there is still significant. We have also observed many comets on Mercury-crossing orbits. On this basis you would not expect it to be safe from the threat of impacts.

Aside from the sheer number of objects that could run into Mercury, there is the issue of their speeds. While the Earth's velocity about the Sun is 18.5 miles (30 kilometers) per second, Mercury rips along at an average of just under 30 miles (48 kilometers) per second. The actual speed varies somewhat because the planet has a fairly egg-shaped orbit, but the overall result is that asteroids strike Mercury at substantially higher speeds, greatly enhancing the energy released on impact.

Because Mercury is so close to the Sun in the sky, ground-based telescopes get a poor view of its surface. Despite this, from the nineteenth century to the first half of the twentieth century, observers claimed to have discerned surface markings and craters on its surface. These now appear to have existed only in the imaginations of the viewers. Not until 1974, when NASA's Mariner 10 satellite first flew by the planet, did space researchers get the chance to see the surface of Mercury in any detail.

Mercury Up Close

The Mariner 10 spacecraft is the only probe to have visited Mercury to date, although three new satellites are planned for launch by NASA, the European Space Agency, and Japan during the next ten years. Mariner 10 made three separate passes as it looped around the Sun. Unfortunately, the coincidence that Mercury's day is precisely two-thirds of its year meant that the probe's camera got much the same perspective each time it flew by, and less than half of the surface could be mapped.

The images we have are more than enough to give us a general picture, however. Like the Moon, Mercury is airless and so it is unprotected from the impactor flux. It even demonstrates the same sort of temporary sodium and potassium clouds detected on the Moon, due to innumerable meteoroid strikes.

There are craters all over Mercury, although not as many per unit area as on the Moon. Why this relative scarcity? Is it because there are intrinsically fewer impacts, or could there be some other explanation? Our best guess is that some major event resurfaced Mercury in the past – either a phenomenal giant impact causing ejecta to cover most of the primordial craters, or perhaps global volcanism spewed basalt over much of the surface. In either case, it is clear that Mercury's surface is dominated by craters resulting from cosmic impacts.

A Matter of Gravity

There is another difference between the craters on the Moon and Mercury, and it stems from the greater surface gravity of the latter, which is almost two and a half times as high as the lunar equivalent. For its size Mercury is very dense – 75 percent of its interior seems to be made of iron – producing this strong gravity. When a projectile excavates a crater, some of the material within is thrown out into space, perhaps to escape, but most of it is scattered sideways across the surrounding surface. The distance that ejecta will fly before landing depends largely on the local gravity. Because the gravitational attraction of Mercury is so high, the displaced rocks do not fly far, and the ejecta blankets around Mercurian craters are much more compact than their lunar counterparts.

Ice in the Shadows

There is another way to study Mercury, and that is by using powerful radar systems. Radar echoes were first received from the planet in 1964, and these indicated its rotation rate and explained its day length as mentioned above. Improved technology led to a surprise in 1991: radar images of Mercury indicated strong returns from the polar regions, identifying reflections from some sort of electrically conducting material in the floors of craters located there. One possible explanation was the presence of ice. But how could frozen water persist on a planet so near to the Sun, and thus so hot?

The answer is startlingly simple: sunlight never penetrates these polar craters. Unlike the Earth, whose spin axis is tilted by some 23.4 degrees, Mercury spins with its axis very close to perpendicular to its orbital plane. This means that the ice is perpetually shaded from direct sunlight by the elevated rims. Since Mercury lacks any significant atmosphere, there is no way for heat to be transported into the craters by any other means. Calculations indicate that the ice will be stable for millions of years. But how did the ice get there? Once again, impacts seem to hold the key. Comets are largely made of ice, and it appears that Mercury manages to hold on to a small fraction of the frozen water from the projectiles that perpetually bombard it.

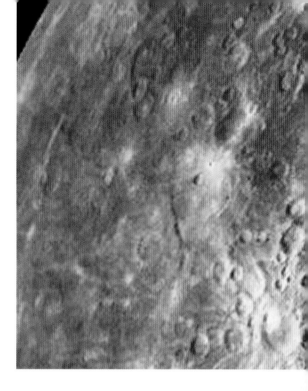

Above This color-coded image of a section of Mercury gives a good idea of which craters are old and which are young. The bright yellow blotch is crater Kuiper with its obviously fresh ejecta blanket.

Venus Reveals All

Moving outward from the Sun, the next planet is Venus. But she keeps her face hidden.

On the way to Mercury, Mariner 10 flew past Venus and returned images similar to those obtained by several preceding U.S. and Russian satellites. These images were pretty uninteresting: little could be seen because of the dense clouds that perpetually shroud the planet.

The Perpetual Veil

This was not a surprise. We had known for some time that the high reflectivity of Venus's clouds makes our nearest planetary companion one of the brightest objects in the sky when seen as the Morning or Evening Star at different times of year.

Venus's opaque cloak, however, is quite unlike the familiar water-vapor clouds of Earth. The droplets that make up the Venusian clouds consist largely of sulfuric acid, with various other noxious chemicals added for good measure.

One might have wondered whether Venus had a solid surface at all, or was it merely a ball of gas and fluid like the outer planets. Initial Earth-based radar soundings in the 1960s showed that Venus does indeed have a solid, rocky surface some tens of miles below the cloud tops. But from such a distance, these radars could show little detail. In our quest for craters, we would need to wait for a closer look.

The atmosphere of Venus is known to be extremely dense and extremely hot. Down on the surface of the planet, the pressure is ninety times that of the terrestrial atmosphere, and the

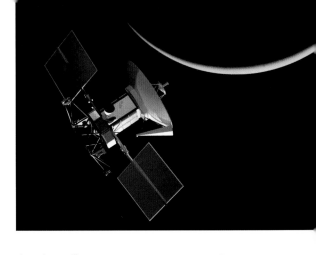

Above NASA's Magellan satellite carried a powerful radar system, allowing the surface of Venus to be mapped with unprecedented detail, despite the clouds that shroud the planet.

Above This color-coded Magellan radar image of the surface of Venus reveals the existence of impact craters and their accompanying ejecta blankets.
Right The surface of Venus is rocky and chopped up largely by the effect of impacts, as shown by these pictures obtained by the Soviet Venera 14 lander.

ВЕНЕРА-14 ОБРАБОТКА ИППИ АН СССР И ЦДКС

ВЕНЕРА-14 ОБРАБОТКА ИППИ АН СССР И ЦДКС

CRATERS, CRATERS, EVERYWHERE

temperature around 900°F (500°C). The atmosphere is predominantly carbon dioxide – indeed it was through studies of how the elevated temperature of Venus comes about (since it is shrouded in clouds, Venus actually absorbs less sunlight than Earth) that scientists gained a better understanding of the greenhouse effect at work on our own planet.

The Moon and Mercury have many impact craters, but they have no atmosphere to shield them. Could the thick Venusian gaseous layers protect that planet from impacts?

Radar Mapping

To see through the clouds you need to use radar, preferably from up close. At the end of the 1970s NASA sent its Pioneer Venus Orbiter to circuit the planet, equipped with radar equipment that mapped the elevations of the surface far below the clouds. The heights of the terrain were derived from the delay times of the echoes bounced off the ground, but the strength of the returning signals also gave information about the texture of the surface. Was it smooth, flat rock, laid down as a lava flow? Or was it rough and blocky, shattered by asteroid impacts? The evidence was clear that a chopped-up, impact-produced surface was the norm.

Two Russian probes, Venera 15 and 16, soon followed in radar mapping parts of Venus, but the big step forward came when NASA's Magellan spacecraft orbited between 1992 and 1994. On board was a radar system that gave much better detail than previous probes, and it traced 98 percent of the surface of Venus to a resolution as good as a city block.

Did Magellan find craters? You bet. Some 1,000 impact craters have been identified to date, all larger than 2 miles (3 kilometers) in diameter. The size is significant. It is much greater than the limiting resolution of the radar, so it is not a question of smaller craters being missed due to limitations of the data. Rather, it is clear evidence that the thick Venusian atmosphere does have a shielding effect. Cosmic projectiles smaller than a few hundred yards in size – which would produce craters a mile or two in dimension if they were to strike the surface unimpeded – are broken up in the atmosphere, dissipating their tremendous energies far above the surface. They produce blast waves – and there is evidence of such blasts shaping some areas of Venus's surface in the radar imagery – but no small craters. Only the big rocks get through, leaving their effects for all to see.

Below The rough, blocky nature of the ejected material from impact craters make them appear bright in radar images such as this oblique view built up from Magellan data.

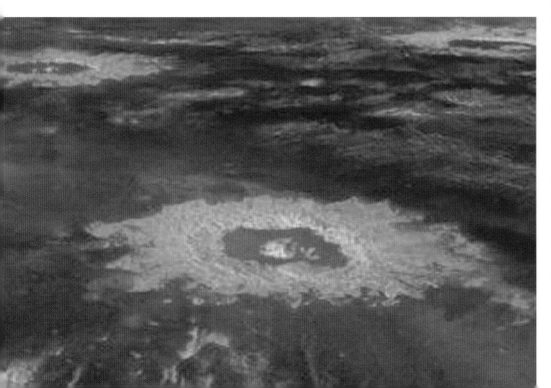

Mars Attacked

Spacecraft imagery of Mars has shown that it too is covered with impact craters.

As seen from the Earth, Mars is a reddish color, with white polar caps, and dark blotches dotted across its surface. Nineteenth-century astronomers viewed Mars through their telescopes and formed a variety of imaginative theories based on their observations. Some held the belief that the planet's surface was covered in vegetation and contained a network of rivers. Others took this idea even further and claimed that what they saw was evidence of intelligent life on Mars. But as far as we know there is no life there. Whether there ever has been – even a life form as simple as algae – is another question.

The debate and confusion surrounding the appearance of Mars that began a century ago continued right up to the space age. The problem was that the planet is just too far away for ground-based telescopes to resolve many of its details. Even today the images retrieved using the Hubble Space Telescope, in orbit just above our own atmosphere, do not provide clear evidence of the nature of the planet's surface.

During the 1950s, astronomers who knew how the lunar craters were formed reasoned that asteroids and comets must also have hit Mars. Their hunch was confirmed when the first probes reached the planet in the 1960s and relayed images that showed a plethora of craters dotting its terrain.

Hit Below the Belt

Although the numbers of comets crossing each planetary orbit increase as you step outward in the solar system, this is a minor consideration in terms of the relative numbers of impacts on Mars and Earth. There may be twice as many comets crossing Mars's path as there are crossing Earth's, but the greater gravitational field of our planet leads to an increase in the impact rate per unit area. The effects cancel each other out, to a large extent.

Below This panorama obtained by the Mars Pathfinder lander shows the blocky surface of Mars and its pinky-brown sky. The Sojourner rover vehicle is next to the boulder in the foreground.

Above A detailed view of Mars reveals many craters. The largest here is Schiaparelli, named after the nineteenth-century Italian astronomer who thought he saw rivers on the planet. At lower right, white carbon dioxide frost can be seen in a vast basin called Hellas. **Right** Canyons such as this indicate that water once flowed on the surface of Mars. Note the pervasive scattering of craters.

CRATERS, CRATERS, EVERYWHERE

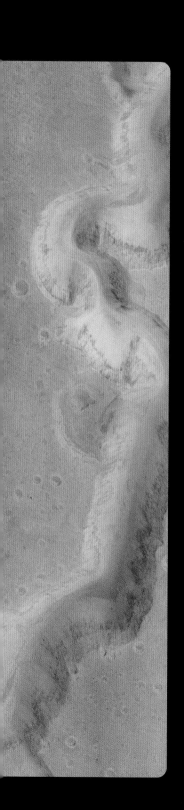

A far greater influence on the expected hit rate for the red planet is that there are many more asteroids on Mars-crossing orbits. Mars lies just below the inner edge of the main belt of asteroids, and it is much easier for one of those asteroids to be pulled into a Mars-crossing orbit than it is to produce an elongated orbit that reaches inward as far as Earth. There seem to be at least ten times more potentially Mars-colliding asteroids than Earth-colliding asteroids.

Eroding the Evidence

Although Mars does not possess a thick atmosphere like Venus, or even Earth, it does have a gaseous cocoon that absorbs the smaller meteoroids and dust grains that come its way.

Mars also has water, but mostly it is locked away, frozen in the polar caps and subsurface permafrost. In the past Mars was warmer, and rivers did indeed flow on its surface. In that epoch many of the impact craters formed on Mars would have rapidly eroded away, leaving a surface that seems relatively uncratered compared to our Moon, where the only event that can wipe out a crater is the formation of a new one by another impact.

Other agents of erosion have also been at work on Mars. Huge volcanoes dot its surface, and there is evidence of widespread lava flows. These must have covered many of the craters that were excavated early in the history of the solar system – when the projectile flux was far higher than it is now. Evidence suggests that there has been recent volcanic activity on Mars (recent, that is, relative to the overall age of the planet) although there are no signs of eruptions happening at the moment.

Another important agent of erosion is the Martian atmosphere. It can lay down thick layers of frost, and whip up ferocious dust storms, chipping away at craters and covering them over with dust and sand.

Right While large craters have central uplifts formed by rebounding target rock – just like dropping a sugar cube into a cup of coffee – smaller craters tend to be simple bowl shapes, like this Martian example. Surrounding it lies a pancake of ejected material.

Martian Meteorites

Many will remember the controversy that arose following NASA's 1996 announcement that it had found evidence for past microbial life on Mars. This claim was based on the alleged discovery of microfossils inside a meteorite known to have come from Mars. Although few scientists believe the fossil interpretation, we are quite certain that this meteorite and more than a dozen others are samples of the Martian surface. How did they get to Earth?

The answer lies in the devastating energies released by cosmic impacts. When an asteroid or comet excavates a crater, the rock and other ejecta have to go somewhere. Most of the material is retained by the planet's gravity and ends up scattered over its surface, especially near the impact site. But if the collision is extremely energetic, some target rocks may have ejection speeds sufficient to leave the planet. These escaped rocks travel through interplanetary space, typically for millions of years, until they run into something that stops them – like the Earth.

Apart from the Martian meteorites, we also have several lunar meteorites – free samples of the Moon that were sent to us by impact events. The likelihood is that there are millions more of these pieces of the Moon and Mars spread all over the globe.

Jovian Thunderbolts

● **Proof of the phenomenal energy released in a major impact was obtained when a string of comet fragments slammed spectacularly into Jupiter.**

In July 1994, the eyes of the astronomical world, and the media too, were turned toward just one celestial object: Jupiter. We knew that the king of the planets was set to take a series of sucker punches as a chain of comets careered into it. Anyone who imagined that havoc-wreaking impacts were things of the distant past was about to receive a nasty shock. The Jovian thunderbolts produced by Periodic Comet Shoemaker-Levy 9 (SL9) proved the notion that all is now safe in the universe to be utterly false.

Eugene and Carolyn Shoemaker, together with David Levy, formed a prolific team of comet and asteroid searchers. They discovered many Earth-crossing asteroids, and a host of comets bear their names. But one — SL9, pictured on page 26 — gained more fame than all the rest.

The String of Pearls

In March 1993, the Shoemaker-Levy team found SL9 to be a celestial phenomenon unlike any they had seen before. While looking at

Right After the impacts: the dark scars in the southern hemisphere are obvious. The black spot in the north is the shadow of one of Jupiter's moons.

Above The impact regions, spaced out due to Jupiter's rotation, continued to glow for days after each event, as shown in this infrared image.
Left The flash of an impact event, an explosion releasing far more energy than the most powerful nuclear weapon.

photographs of the sky close to Jupiter, they spotted a bar-shaped image with a wide tail, similar to the profile of a normal comet but spread curiously sideways. Other astronomers were soon able to confirm their discovery, and with the better resolution afforded by larger telescopes and electronic detectors, the reason for the bizarre spectacle soon became clear.

SL9 was a broken-up comet. It is known that weak objects passing close to a planet will be pulled into their constituent parts by the tidal force they experience during their passage. The rings of Saturn seem to have been formed in this way, when a large comet — maybe more than one — happened to pass by that planet. This was clearly what had happened to SL9.

Subsequent observations confirmed that SL9, unlike other comets that orbit the Sun, was actually in an orbit around Jupiter that took a few years to complete. No one knows for sure how long it had been there, although it is likely that it was trapped by the giant planet only a few decades ago. Perturbations of its path must have gradually dropped its minimum distance above the Jovian cloud tops, so that in 1992 it passed too close and was broken into at least

CRATERS, CRATERS, EVERYWHERE

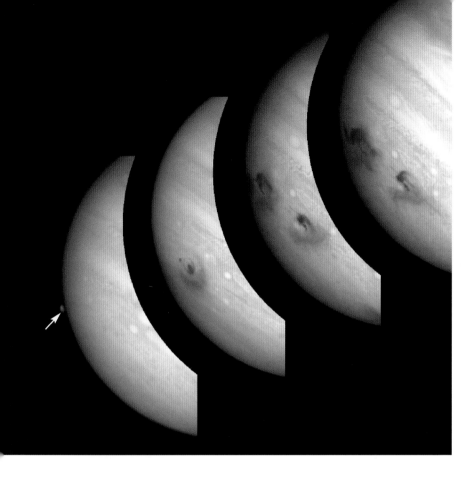

Left The changing shape of the dust cloud formed in Jupiter's atmosphere by one of the collisions is depicted in this sequence, taken over several days. It starts with the impact and its plume rising above the edge of the planet, shown here far left. Even months later, the residual scars could still be seen.

gas giants because they are, in effect, giant balls of gas. The bright disk you see in a telescope is simply the cloud tops. Going deeper below that level, the pressure grows until the major constituents, hydrogen, and helium in particular, liquefy. But the solid cores of the planets Jupiter, Saturn, Uranus, and Neptune can be only small fractions of their overall bulks. There are no rocky surfaces to be seen. Jupiter simply swallowed up the massive comet fragments.

This does not mean that there was no blast. Far from it. In the months leading up to "impact week" in mid-1994, space researchers debated whether anything at all would be seen. But what was observed exceeded every expectation. The incoming SL9 fragments were of various sizes, with the bigger ones measuring over a half mile (800 meters) across, according to our best estimates. They slammed into the planet at just below Jupiter's escape speed (around 38 miles or 60 kilometers per second) where they encountered the frictional force of the atmosphere trying to slow them down. The enormous shock caused them to explode, the larger components liberating energy equivalent to about 50,000 megatons of TNT. These were big bangs in anyone's language.

The fireworks show seen from Earth, and from the Galileo space probe, was spectacular. Phenomenal flashes emanated from the impact zones, and as the dust settled – quite literally – darkened regions appeared that measured up to four times the area of the Earth. These persisted for months after the event. Jupiter had been given a series of black eyes. Thankfully it hadn't been our own planet in the line of fire.

Above Although the SL9 impacts were on the dark side of Jupiter, normally hidden from view, the Galileo space probe, en route to the planet, was able to peek around the corner and record what happened. This image shows – as a bright white flash at left – one of the explosions that occurred.

twenty major fragments, and myriad smaller ones. Astronomers described it as resembling a string of pearls.

Its disintegration meant that large amounts of dust were liberated into space, scattering sunlight back to telescopes on Earth, and allowing solar heat to reach the volatile ices previously hidden within the comet's interior. These factors served to brighten it considerably.

By the end of 1993 the initial suspicion – that the chain of newborn daughter comets would run into Jupiter on the next approach – was confirmed. We were going to see not just one, but a whole series of impacts upon a planet.

Black Eyes All Round

What we did not see, though, were any craters being formed. The outer planets are termed the

It's Happening Now

It's not just the planets that get hit: recent satellite observations have demonstrated that comets frequently collide with the Sun.

Below By blanking out the Sun using a screen, the coronagraph on board the SOHO (Solar and Heliospheric Observatory) satellite produces an artificial eclipse, allowing the streamlines of the solar corona to be detected. This also makes it possible to detect comets plunging into the Sun, as in this example. **Right** Produced using ultraviolet satellite data, this sequence of images shows a comet falling toward the Sun in February 2000.

One lesson learned from the cometary impacts on Jupiter in 1994 is that such events are not restricted to the past. They are happening *now*. Since those Jovian impacts, dozens of comets have struck another solar system object: the Sun.

Comets are often seen to split apart in interplanetary space for no apparent reason, although we are able to hazard a guess at the underlying causes. Comets seem to be quite fragile. As they are warmed by solar radiation, their constituent ices evaporate. Lacking enough self-gravity to hold them together, they may be torn apart by thermal stresses and gas pressure. The most famous example of this in recent centuries was Comet Biela, which separated into at least four distinct nuclei in the 1840s. The orbit of the resulting debris cloud evolved in such a way that a few decades later it came around to intersect the Earth. This produced several phenomenal meteor storms, the most prominent in 1872. Thankfully none of the larger lumps struck home.

Ancient records also describe the splitting of large, bright comets. One of the best-known occurrences was in 372 B.C., when Greek historian Ephorus reported that a comet had separated into two parts. This is significant because, in the nineteenth century, astronomers were able to link that comet with a whole series of others that had been seen more recently. These comets were peculiar because they all shared similar orbital characteristics: retrograde, long orbital periods (hundreds or thousands of years), all with a very small perihelion distance so that they almost skim the solar surface each time they return. These became known as the Kreutz group, after the German astronomer who investigated them.

The theory runs that all of these separate comets are components of a former giant

Right This comet fell into the Sun in late December 1996, and has since been known as the Christmas Comet. The Milky Way is seen in the background.
Background Another view of the Christmas Comet and its extensive tail.
Center, top to bottom This four-hour sequence from SOHO shows not just one but two comets approaching the Sun. The size of the Sun itself is shown by the white circle drawn on the screen at the center.

progenitor that split apart when passing close by the Sun. The physical cause of its break up might have been the same type of tidal force that fragmented Shoemaker-Levy 9, or alternatively the incredible thermal stress that a comet must experience coming so near to the solar inferno.

Satellite Observations

Up to the 1970s several more comets in this group were accumulated in our data banks. These were large fragments of the original, making them relatively bright. They could be seen as they passed the terrestrial orbit on their way toward the Sun, and again on their way back – if they survived. But the space age brought a new type of observation.

In the past, solar physicists who wanted to study the Sun's corona or chromosphere (the tenuous upper layers of the solar atmosphere) had to chase infrequent total eclipses. Only when the Moon blanked out the bright photosphere below could they get a clear view of their subject. Now, though, it is possible to equip a satellite with an instrument that produces an artificial eclipse. This device employs an opaque disk in front of the telescope to perform the same function as the Moon, enabling the behavior of the solar atmosphere to be monitored continuously.

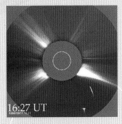

16:27 UT

16:58 UT

17:56 UT

20:13 UT

The images sent back by the instruments on board Solwind and SMM (or Solar Maximum Mission) produced a real surprise, though. They showed a succession of tiny comets, typically about the size of an average house, falling into the Sun. These comets, most of which seemed to be members of the Kreutz group, could not be detected from the ground because they were too faint, and so they were drowned out by the solar glare.

The number of these comets observed has risen in the past few years thanks to the Solar and Heliospheric Observatory (SOHO) satellite. Dozens of them have been witnessed diving suicidally into the Sun. Clearly the progenitor comet must have been huge.

The Sun is struck repeatedly by comets. Could this string also hit the Earth? In principle, a cluster of cometary fragments like the Kreutz group crossing the terrestrial orbit would be a very dangerous proposition. The good news is that because of the orientation of their trajectories, none of this particular group can run into planet Earth. But there are other candidates whose paths are much less reassuring.

The Moons of Jupiter

The effects of a comet strike on a gas giant disappear in months or years. But on Jupiter's large moons the evidence of impacts is preserved for all to see.

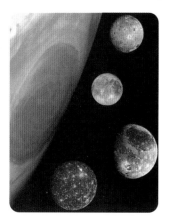

Above The four Galilean moons of Jupiter have surfaces shaped to differing extents by impacts.

The huge explosions that accompanied the "string of pearls" impact on Jupiter were each equivalent to a nation's entire nuclear arsenal being detonated simultaneously in one place. The aftermath – vast black scars in the cloud decks, far larger than the whole of the Earth – was visible for months, traces still being detectable years later. But as we have seen, Jupiter has no solid surface, and therefore no craters.

To study the long-term impact rate of meteorites and comets, you need a solid planet capable of preserving evidence of the insults it has suffered. Fortunately Jupiter is orbited by its own system of mini-planets, four of them as big as, or bigger than, our own Moon. These are the Galilean satellites, and each tells us something different about cratering in the solar system.

Ganymede and Europa

Although the Jovian system was briefly visited by the Pioneer 10 and 11 and Voyager 1 and 2 space probes in the 1970s, the era of detailed study of the moons of Jupiter did not begin until December 1995, when the Galileo spacecraft became the first artificial satellite of that planet. That probe has returned excellent imagery of the surfaces of the Galilean satellites, enabling us

to count their craters and make deductions about their histories.

Crater distributions are of interest to planetary scientists not just because they want to know how often impacts occur. Craters also provide a dating system: a surface bearing few craters must be younger than one that is heavily pockmarked. In this respect, Ganymede and Europa provide a fascinating contrast.

Europa's surface consists of a deep crust of ice floating on an immense ocean below. Such a structure, even many miles thick, cannot preserve craters for as long as solid rock. The face of Europa is dynamic, and so we see comparatively few craters on its surface.

Ganymede, a mixture of about 60 percent rock and 40 percent ice, is much more stable. Impact craters formed on Ganymede last far longer than those on Europa, resulting in a much more densely cratered terrain. Modest-sized projectiles produce the small conventional craters that dot Ganymede's surface. But the larger impacts may leave very different structures. A big comet or asteroid slamming into Ganymede punches through its surface layers and reaches the mixture of ice and rock below. This slushy sub-surface material is not strong enough to retain the

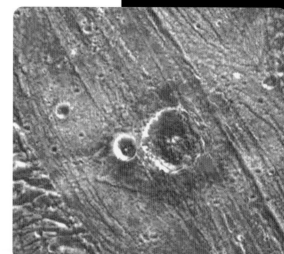

Above A false color rendering of Callisto shows clearly the locations of its craters.
Below The Nergal area on Ganymede showing a pair of craters that may have been formed at the same time.

CRATERS, CRATERS, EVERYWHERE

Left The shape of the Pwyll crater on Europa is clear in this topographic mapping.

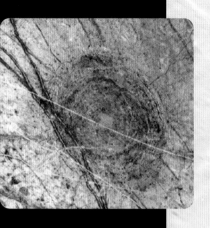

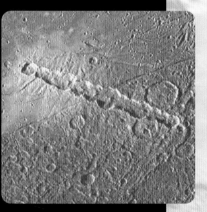

Top The icy surface of Europa, marked by cracks, carries many impact craters of peculiar form.
Above A crater chain on Ganymede, formed by a string of projectiles hitting this moon as they passed by Jupiter.

shape of a deep crater bowl, so it fills the depression. The end result is called a palimpsest. Many of these blotches are seen on Ganymede, clearly testament to a past when the moon was not as cold and solid as it is now.

Chains on Callisto and Ganymede

Callisto is similar to Ganymede, but deserves special mention because it preserves evidence that episodes similar to the Shoemaker-Levy 9 break-up have occurred numerous times in the past. Imagine that, instead of having an unobstructed route into Jupiter's clouds, SL9 had encountered one of the Galilean moons in its way. A chain of impact craters would have resulted. Voyager imagery from 1979 had shown such chains on Callisto, but they remained a puzzle until the discovery of SL9 gave us the clue to their origin. More than a dozen such chains are known on Callisto, and Ganymede also has several unmistakable lines of craters.

Churning Up Io

The innermost of the Galilean moons is Io. Although it is not so close to the giant planet that it runs the risk of being ripped apart like SL9, it is near enough for tidal forces imposed by Jupiter's immense gravity to cause perpetual flexing and compression. The resulting friction generates heat, and the end result is that Io is very hot, with a mostly molten interior. This manifests itself in the extreme volcanic activity that moon displays.

There are craters on Io, but they are volcanic craters. More than eighty active volcanoes have been charted, all spewing out material that regularly recoats its entire surface. Consequently, Io has the youngest surface of any solar system body, and we do not expect to discover any impact craters there.

This is a matter of more than academic interest, and it provides a useful lesson with regard to the Earth's cratering history. Statistically speaking, Io must have been struck as many times as its close neighbors, though we see few impact craters. Geomorphic activity, on Io or on Earth, can result in the complete obliteration of impact craters. We should not be fooled. Absence of evidence is not evidence of absence.

To Saturn and Beyond

Plenty of impacts have occurred in the outer solar system. Nowhere is safe from the threat.

Things slow down as you move out through the solar system. The speeds required by an object to remain in orbit around the Sun drop, from 18.5 miles (29.8 kilometers) per second for the Earth to 8.1 (13.0) for Jupiter, and then 6.0 (9.7), 4.2 (6.8), and 3.4 (5.5) for Saturn, Uranus, and Neptune, respectively. You might imagine that the impact speeds drop off in the same way, but you would be wrong.

subject to an intense flux of impactors. Whether icy or rocky, the satellites of Saturn, Uranus, and Neptune are festooned with craters.

There is one exception. After Ganymede, the largest moon in the solar system is Saturn's Titan. Like Venus, Titan's surface is shrouded in perpetual clouds, although in this case they are murky affairs composed largely of methane droplets. On Earth we'd call it smog: it's the same orange-brown color known to many city dwellers. Because of this smog, the Voyager spacecraft were unable to detect the surface of Titan as they flew by in the early 1980s. We will learn more about the properties of Titan, including perhaps whether there are craters below the clouds, when the Cassini probe reaches Saturn in 2004.

Above Saturn's moon, Mimas, has a profile dominated by one large crater about 80 miles (130 kilometers) across. **Left** Saturn is a gas giant, showing no solid surface to preserve craters, but its moons are densely pockmarked by past collisions. **Below** Ariel, one of the icy moons of Uranus, shows many craters despite the impermanence of its surface.

Blinking Iapetus

Iapetus, another of Saturn's moons, has intrigued astronomers for centuries, because it seems to blink at us. When this moon is to one side of Saturn, and therefore coming toward us, it looks dark. Half an orbit later, as it recedes, it appears much brighter. Sir Arthur C. Clarke used this fact in his story for *2001: A Space Odyssey*, interpreting the contrast as evidence for extraterrestrial intelligence.

Now that we have close-up images of Iapetus, we know why it seems to blink. The moon spins on its axis at exactly the same rate as it orbits Saturn, meaning that it keeps the same face pointing forward at all times. This is called tidal locking. Farther out a smaller moon, Phoebe, happens to orbit Saturn the "wrong" way, in a

Any asteroid or comet coming too close to one of the giant planets is accelerated inward by its strong gravity, resulting in substantial impact velocities, typically several times higher than the speeds given above. The outer planets act like cosmic vacuum cleaners, sucking in projectiles too slow to make an escape. The result is that the moon systems of the outer planets are

Right Triton may
have been formed in a
retrograde orbit about
Neptune as the result of
a cataclysmic collision
early in the history of
the solar system.

retrograde sense. This means that Phoebe is often struck by debris on its way past the planet, and the material excavated from its craters tends to spiral in toward Saturn. Phoebe is dark, and so is its ejected material. As it moves inward, a fraction of this accumulates on the leading face of Iapetus, producing that moon's dark/bright blinking effect. Once again, impacts provide the solution to another long-term astronomical puzzle.

Smacking Shakespeare's Satellites

Even before Voyager 2 reached Uranus in 1986, we knew of five large moons orbiting that planet. The photos sent back by the space probe led to the identification of many more smaller satellites. The original five are named for characters in William Shakespeare's play *A Midsummer Night's Dream*. None has escaped the violence represented by random impacts.

A similar story can be told about Neptune. Its moons display the evidence of impacts occurring over the eons. Triton, largest of the Neptunian moons, bears few craters because of its icy nature. Voyager 2 images showed vast ice volcanoes jetting out material that would fall again on Triton's surface, deleting the scars left by

Above Iapetus has a leading face much darker than its trailing side due to accumulating debris as it orbits Saturn.

impinging comets and similar matter. But Triton may tell us something else about cosmic impacts.

Triton, a Space Oddity

The orbit of Triton around Neptune is retrograde, distinctly odd for a body of its size. Phoebe is small, as are the various retrograde satellites of Jupiter and Uranus, which are thought to be merely captured wanderers. Triton is huge by comparison, almost 1,700 miles (2,700 kilometers) across.

Our best guess at how Triton managed to get into a backward orbit is that early on in the history of the solar system, two massive bodies collided near Neptune, the accumulated debris being trapped to form the beast we see now. We may never know for certain, but it seems likely that a phenomenal impact was the genesis of retrograde Triton.

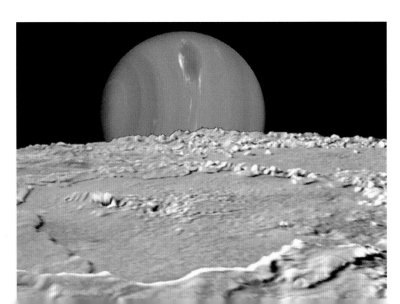

Left Neptune as it would appear from just above the surface of Triton. The surface of this moon is steadily recoated through ice volcanoes spewing out their products, and so few craters last long.

Asteroid versus Asteroid

Asteroids not only slam into planets and moons. They also collide with each other.

This has been known for many decades. The evidence lies in the families of main belt asteroids with very similar orbits. The idea is that large minor planets are shattered into fragments – daughter asteroids – when struck by other wayward bodies. The daughters then continue on similar paths to the parent, but the slight initial variations in trajectory lead to a spreading out in space. So although they appear at different places in the sky, their common origin becomes clear as soon as their orbital parameters are determined.

Vesta is the largest minor planet in one asteroid family, one to which various meteorites studied by researchers seem to be linked. Our guess is that these meteorites started life as the smaller debris liberated in the collision that engendered the family to which Vesta belongs. Their smaller size meant that they had higher relative speeds, and this in turn increased their chances of escaping from their position in the main belt between Mars and Jupiter and evolving Earth-crossing orbits. Eventually the rocks reached the Earth where they were collected by meteorite hunters.

The View from Hubble

From the ground, even the biggest telescopes cannot render the details of an asteroid's surface. The Hubble Space Telescope, though, can be used to get high-resolution images of the largest asteroids. These show evidence of impacts and tell us that asteroids may be both the attackers

and the attacked. This fits in well with our notion that the solar system is anything but a benign place.

When pointed at Vesta, Hubble showed the presence of an enormous crater, 300 miles (480 kilometers) wide and 8 miles (12 kilometers) deep. In fact, the crater is almost as wide as the whole asteroid, as if a large chunk had been sheered off in a cataclysmic collision. The large fragments from this impact are the members of Vesta's asteroid family detected in the main belt.

Getting Up Close

Our best images of asteroid surfaces have been obtained from spacecraft passing close by. During its cruise to Jupiter in the first half of the 1990s, the Galileo probe obtained pictures of two main belt asteroids, 951 Gaspra and 243 Ida. As predicted, their surfaces were covered with craters. No surprises there: we'd grown to

Above Main belt asteroid 243 Ida was photographed by the Galileo spacecraft on its way to Jupiter. It is about 35 miles (56 kilometers) long. A surprise was the discovery of a one-mile (1.6 kilometer) moon of this asteroid, seen to the right of this caption, which has been named Dactyl.

Right Using the Hubble Space Telescope, astronomers have mapped the surface profile of asteroid 4 Vesta, which is 320 miles (520 kilometers) in size.

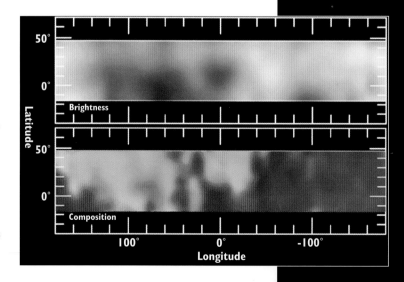

CRATERS, CRATERS, EVERYWHERE

Sequence left
Although an asteroid may simply be cratered by smaller impacts, a large enough collider will shatter it into pieces. In many cases, most fragments will collect together again into a rubble pile held together by self-gravity, but a really powerful collision could spread the debris far and wide.
Right The Earth gets hit too. This is the 1.5-mile (2.5-kilometer) Roter Kamm crater in Namibia, seen from the space shuttle.

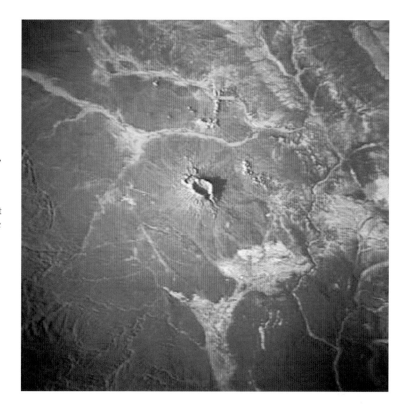

expect every solid surface we encounter to be splattered with impact evidence.

Next to be visited was 253 Mathilde, which was passed in mid-1997 by the Near-Earth Asteroid Rendezvous probe on its circuitous route to an eventual meeting with 433 Eros in February 2000. Again, Mathilde displayed a highly cratered profile. Asteroids do slam into asteroids.

Everywhere You Look

We've just taken a tour of the solar system. Like an ornithologist visiting distant lands in search of exotic birds, we've been interested in only one topic: impact craters. We've found them everywhere we've looked. So why would anyone imagine that the Earth is different? Why should our planet be safe from such random and spasmodic cosmic violence?

Above An aerial view of the water-filled New Quebec or Chubb crater in Canada's north. It is 2.1 miles (3.4 kilometers) in diameter.

Target Earth

Throughout the solar system we've seen that solid surfaces preserve the record of cataclysmic impacts by asteroids and comets. These impacts release energies that make nuclear weapons look like mere firecrackers. If you see a crater 20 miles (32 kilometers) across, you're talking about a 1-million-megaton explosion — over 10,000 times bigger than the largest hydrogen bomb ever tested.

The Moon's pockmarked face bears witness to countless past impacts. The Earth orbits the Sun within the same cosmic hailstorm as the Moon, so we would expect our planet to show similar evidence of damage. More so, in fact, because the Earth's stronger gravity attracts many projectiles inward that would otherwise miss the bull's-eye.

So why isn't our landscape dotted with obvious craters? The answer – that it is, if you know how to look – wasn't widely appreciated until comparatively recently. People simply didn't realize that massive asteroids and comets plummet to Earth from time to time, because such events occur at intervals incomparably longer than a human lifetime. That means, thankfully, that you're unlikely to witness one. But it also means that acceptance of the significance of cosmic impacts has been a long time coming.

Mapping Our Impact History

The number of proven impact craters on our planet has grown rapidly in recent years. There are several reasons for this. People no longer laugh when a geologist suggests that a given structure was produced by a natural bomb from space. The geological community has recognized that impacts leave rock structures that cannot be produced in any other way. Minerals such as stishovite and coesite, formed from shocked quartz, are good indicators of impact origin. Rocks may also

be cracked in a certain way that produces shatter cones, the orientation of the cones indicating the direction from which the shock wave emanated. These too are unambiguous.

But these are scientific matters. Other factors – some of which may seem fairly haphazard – have contributed to an increase in the number of identified impact craters over the past decade or so. The growth in airline traffic, together with the use of automatic pilots that allow the crew on long-distance flights some leisure time to look out the window, has led to many reports of suspicious circular structures on the ground. These sightings are then followed up by scientists, who are able to confirm or deny their impact origins. Similarly, the availability of high-resolution satellite imagery has allowed detailed crater searches to be made.

Incomplete Record

Several previously unknown craters are identified and added to our data banks every year. It follows that there are many others waiting to be discovered.

If you look at the map of known craters you will see concentrations in a few main areas: North America, which is geologically stable and has an educated public. Europe, especially around the Baltic, where the same factors apply. The north and west regions of Australia, where the population density is low but there is little vegetation, so impact scars are laid bare. It is easier to understand why craters are known in these regions, especially when you compare them with the tropics: few craters are plotted there, because they tend to be covered with dense foliage and quickly eroded by heavy rains.

There is no reason to expect that any one region of the planet should receive more impacts than another. It follows that the maximum crater density we find anywhere is a measure of the minimum for the whole planet: the highest number we find can only increase as more craters are recognized, and we can conclude that every region will have at least this level of cratering.

Left Over 200 impact craters have been identified around the world. Where next?

Above left An artist's impression of a twin-lobed or binary asteroid. Separation of the two components just prior to impact would produce twin craters of a kind observed on Earth and elsewhere.

Above The Clearwater Lakes in Canada constitute a pair of craters formed by a binary asteroid strike. The larger scar, which has an uplifted ring within it, is 20 miles (32 kilometers) across; the smaller is 14 miles (22 kilometers) wide.

Meteor Crater

There's no doubt that the best-known impact crater on Earth is Meteor Crater in Arizona.

It is properly called the Barringer Meteorite Crater, after Daniel Barringer, who was one of the first to identify it – back in the early 1900s – as having an impact origin. It is still owned by his descendants, who run it as a tourist attraction. People traveling cross country often stop to take a look, gawking at the huge hole from its rim.

Two Huge Holes in Northern Arizona

You could say – if you include the Grand Canyon – that there are two great holes in northern Arizona. Meteor Crater is about three-quarters of a mile (1.2 kilometers) wide and several hundred yards deep, making it a formidable sight. Although the Grand Canyon is far bigger, Meteor Crater is impressive in a different way.

When you gaze into the abyss of the Grand Canyon you may see it as evidence of the passage of time, with the Colorado River eroding it away over millions and millions of years. Meteor Crater, on the other hand, represents just a few seconds of cosmic violence. The Colorado River is red because of the detritus it carries, excavating the canyon particle by particle over eons. Meteor Crater was formed 49,000 years ago in an instant when a lump of nickel and iron, only about 40 yards or meters across, slammed into the ground at 10 to 20 miles (16 to 32 kilometers) per second, and sent billions of tons of rock flying.

Right Meteor Crater as seen from the space shuttle, an obvious feature in the parched plateau of northern Arizona.

Drilling for Iron

It was the fragments of the impactor that were left lying around that gave the first clue to the crater's origin. Many large lumps were scattered over the surrounding desert, while smaller flakes can still be picked up in and around the crater by dredging a magnet through the powdery soil.

A century ago, the U.S. Geological Survey classified Meteor Crater as a cryptovolcanic structure: a blowhole from past volcanic activity that had long since ceased. In fact, there is no real evidence of volcanism nearby – the rocks are of sandstone and limestone, which are unrelated to volcanic processes.

Daniel Barringer was outside the scientific mainstream – it is often people in this position who are responsible for truly great advances – and he correctly interpreted the shape of the crater, the metallic debris, and the overthrown rocks around its rim as evidence of an impact. He bought the structure, then known as Coon Butte, and in the end it proved to be a great

asset, although not in the way he anticipated.

Barringer got one thing fundamentally wrong. He thought that the major portion of the projectile must be buried below the crater floor, and he believed that such a vast lump of near-pure iron and nickel would make him rich. He began boring into the floor, and then into the walls of the crater, in search of the metals. But no matter where he dug, he couldn't find the mother lode. What he didn't understand was that a cosmic projectile possesses more than enough energy to vaporize when it rapidly decelerates in an impact with the solid Earth.

All that was left of the massive meteorite were a few scattered shards. The majority had evaporated and was carried away in the huge plume that would have risen high above Arizona. If anyone had been there to watch, it would have looked like the mushroom cloud from a nuclear detonation. After all, we are talking about an explosion that released the energy equivalent of about 20 megatons of TNT.

Above An aerial view of Meteor Crater showing how its rim is raised above the surrounding plain, and how deep the impact tore into the rock. **Inset** Gene Shoemaker, the modern pioneer of Meteor Crater. His investigations established beyond doubt that it was formed by an impact.

Impacts on North America

Meteor Crater is not an isolated phenomenon. Many others are known.

Acceptance by the geological fraternity that there have been major impacts on the Earth did not, in fact, come from Meteor Crater, but from studies of a crater in Odessa, Texas, in the 1920s and 1930s. Like the one in Arizona, the Odessa crater was surrounded by metallic debris from the impactor. It was clear that these were fragments of a meteorite, and not terrestrial iron. The association was undeniable. We are still talking, though, about relatively small craters, structures a mile or two wide that can be appreciated from the ground.

The problem with bigger craters is that they are less obvious when you're standing right on top of them. A crater 20 miles (32 kilometers) in diameter may start out many miles deep, but the ground cannot sustain such a cavity so it slumps to fill in the hole. Standing on one edge, the far rim may be below the horizon. It would take detailed mapping to determine the structure's shape and identify it as an impact crater. Other evidence would be needed to back up the hypothesis: minerals metamorphosed by shock, or melted rocks, or geological charts to show that the rock strata had been folded back over themselves. Finding large craters was difficult as long as we were restricted to the ground. Observing them from airplanes has helped. Looking down from orbit is even better.

Chesapeake Bay

One of the largest craters in North America has been recognized as such only in the last few years, despite the fact that it is right under the noses of the majority of the population. Admittedly it is underwater, but it lies on the

Above An artist's impression of the massive impact that formed the Chesapeake Bay crater 35 million years ago.
Right The Manicouagan reservoir in Quebec occupies a circular impact crater 60 miles (100 kilometers) across. This view was obtained from orbit; lower left is the tail of the space shuttle.

doorstep of the U.S. Naval Academy, a NASA research center, the headquarters of the CIA, the National Security Agency, and the U.S. government itself. It is the southern end of Chesapeake Bay. At 53 miles (85 kilometers) across, there are few larger impact structures known on the planet.

Canadian Craters

Many of the best-known impact craters are found in Canada. Since much of northeastern North America has been geologically stable for millions of years, it has faithfully recorded the traces of impacts. This area is known as the Canadian Shield region. But there are other reasons why Canada is a good place to look for craters. During the summer, large circular lakes hint at a possible impact origin. During the

Clearwater Lakes
Deep Bay
Manicouagan
Chesapeake Bay
Odessa

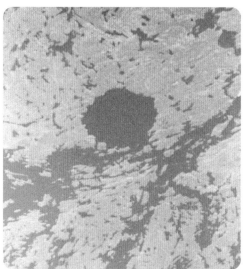

Above Deep Bay is a lake-filled crater in Saskatchewan. It is 8 miles (13 kilometers) wide.

winter, they are frozen and covered with snow, which makes them easier to spot against the darker background of tree-covered terrain.

Twin Craters

Canada's Clearwater Lakes (see page 51) represent something of a puzzle: a pair of impact craters so close together that the asteroids that created them must have landed at the same time. How could that happen? Similar paired craters have been found elsewhere, so we know it's not a fluke. As so often the case in science, a number of realizations happened simultaneously. We saw Comet Shoemaker-Levy 9 split into fragments and slam into Jupiter. We found that some asteroids appear to have satellites, which may be quite large. And radar images showed that some Earth-crossers might be two large lumps revolving in contact. As such a binary asteroid approaches on a collision course with Earth, it would be expected that tidal forces would rip the two bodies apart, causing them to impact at close but distinct points. This seems to be the answer to the riddle of twin craters.

A European Tour

● **A large number of impact craters are known in Europe, especially in the north.**

Like Canada, northern Europe is geologically stable and snow-covered in winter. It has also had a scientifically literate population for an extended period. These are people who ask questions and wonder why a certain lake or set of hills is circular, or strange in appearance. The region known to geologists as Fennoscandia (Finland and the Scandinavian countries) is prolifically cratered, enabling the global impact rate to be better estimated.

Underwater Craters

Not all of these craters are in lake-filled pine forests. Two underwater craters are known in the Baltic Sea. Only two other suboceanic craters are known in the world. All four are on continental shelves; no proven impact craters have been found in the ocean basins. But as 70 percent of the Earth's surface area is under seawater, we may expect to discover many more craters as we learn more about the bottom of the oceans.

Getting back to the Baltic, the following example demonstrates how much more frequently craters are being found today than was the case in the early years. At the beginning of 1994, a major international conference was announced by Swedish and Finnish astronomers to bring together researchers working on various facets of the impact problem. This was to be held in the little town of Mariehamn, on

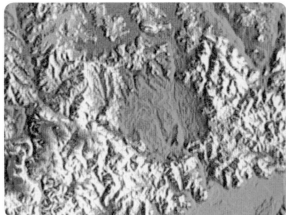

Below The depression of the 15-mile (24-kilometer) Ries crater in Germany is clearly visible in this color-coded topographic map.

Right The Steinheim crater in Germany is about 2.4 miles (3.8 kilometers) wide. Its circular form and central uplift are obvious in this aerial photograph, as are the small towns and farms it now contains.

Finland's Åland Islands. By the time the conference was held in August of that year, it was recognized that Lumparn, the large natural harbor in the midst of the Åland archipelago, was actually itself an impact crater 6 miles (10 kilometers) across. This gave the astronomers a chance to inspect the crater up close.

Some craters were formed millions of years ago and have since filled with fertile soil. They support farms and small towns, as in the cases of Steinheim in Baden-Württemberg, Germany, and the 16-mile (26-kilometer) Ries crater, Germany. The town of Nördlingen resides within its basin.

The Impact Rate

Geologists have calculated a lower limit for the rate that asteroids and comets strike the Earth through a tally of craters — in a variety of sizes — in areas such as the Canadian Shield and Fennoscandia, while taking the age of the target

Right The Mjølnir crater on the ocean floor under the Barents Sea off the north coast of Norway has recently been identified by University of Oslo geologists using seismic sounding. Using such data profiles as shown here, a crater 24 miles (40 kilometers) wide is measured.

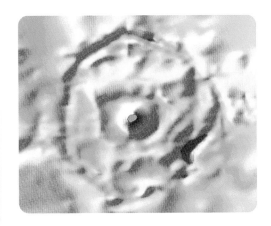

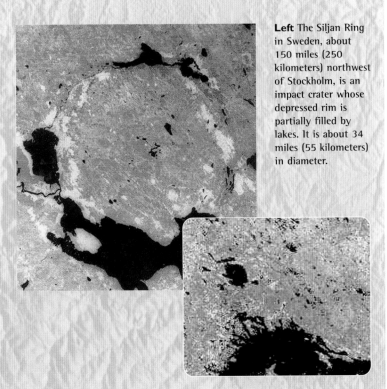

Left The Siljan Ring in Sweden, about 150 miles (250 kilometers) northwest of Stockholm, is an impact crater whose depressed rim is partially filled by lakes. It is about 34 miles (55 kilometers) in diameter.

rock into account. Astronomers are able to derive an impact rate working in the opposite direction. They count up the number of known Earth-crossing objects and calculate a terrestrial impact probability for each. The overall sums give a result from the astronomical perspective.

Considering the uncertainties involved with both methods — How many craters have been missed? How many more asteroids are there to find? — it is amazing that the two methods provide answers that are reasonably consistent. They suggest that a projectile at least a half mile (800 meters) wide strikes our planet about once every 100,000 to 200,000 years.

Left A view from orbit of the Janisjärvi impact crater in Russia, near the border with Finland. The 9-mile (14-kilometer) water-filled crater is the dark, rounded region to the left of center. Below it is the northern fringe of Lake Ladoga, which lies to the north-east of St. Petersburg.

Outback Australia

Wolfe Creek •
Gosses Bluff • • Boxhole
• Henbury
Lake Acraman •

● **In the 1930s, as the impact advocates were gaining a tenuous foothold in the United States, news was filtering through to Australia, then still a frontier country.**

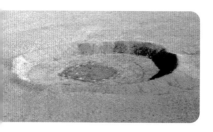

Above Wolfe Creek crater in Western Australia is about 960 yards (875 meters) in diameter.
Below right In this view from the space shuttle, the circular patch at lower left in this photograph is Lake Acraman: a 60-mile (100-kilometer) impact scar formed 570 million years ago.

Above The network of small impact craters at Henbury were produced only a few thousand years ago.

Although Europeans had been in Australia for 150 years, there was a lot of unexplored terrain, scientifically speaking. Those roving through the Outback had no preconceptions about what they might find in this parched land. As a result, impact craters were rapidly identified.

The southeastern part of Australia is largely covered with flood plains, and craters do not last long in that environment. There are no known impact craters in the states of New South Wales or Victoria, and only one suspected impact in the island state of Tasmania. The situation is similar in Queensland, which has substantial vegetative cover and monsoonal weather in the north. The arid areas of South Australia, the Northern Territory, and Western Australia, however, are geologically stable and largely bare of vegetation. This makes them ideal locations to find obvious craters.

Henbury Craters

One of the earliest sites to be identified consists of not just one crater, but a group of around fifteen. The Henbury craters lie about 80 miles (120 kilometers) south of Alice Springs, in the center of the continent. They are not huge: the largest crater is less than 200 yards (180 meters) across. On the other hand, they are young, with an estimated age of around 4,000 years. The asteroid that produced them would have been about the size of a large house, breaking up in the atmosphere and producing a scattered pattern of craters spread out over the area of a football field.

As is the case for all small craters, the projectile was metallic. A rock the size of a

house would not have made it through the atmosphere without vaporizing, and so no crater would have been formed. In fact, even a rock the size of a twenty-story building would be unlikely to reach the ground. All except for one of the known terrestrial craters smaller than a couple of miles in diameter are known to have been caused by metallic asteroids.

At Henbury it is easy to find remnant fragments of the impactor up to a quarter of an inch (6 or 7 millimeters) or so in size. They are scattered all around, and the tiniest pieces can be removed from the soil using a magnet. There are also many obvious pieces of melted rock, which have a glassy appearance. Their darkness contrasts with the lighter dusty soil and gives us a clue to the violence that occurred there.

Northeast of Alice Springs is another crater called Boxhole. This was also formed by a

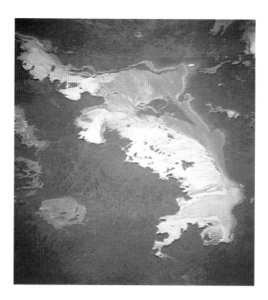

Left The circular structure in the lower part of this space shuttle image is Gosses Bluff. This is only the central part of a much larger impact scar, which has mostly eroded away. To the north lie the MacDonnell Ranges.

metallic impactor, at about the same time as Henbury. It could be that the two are related. Perhaps an asteroid sheared in two, leaving craters a couple of hundred miles apart.

Gosses Bluff and Lake Acraman

Two hundred miles (300 kilometers) to the west of Alice Springs is an entirely different impact scar. Gosses Bluff is a circular ring of hills about 4 miles (6 kilometers) across. This is the result of an impact that occurred about 142 million years ago, but it is not the crater rim. Gosses Bluff is an example of a central uplift as seen in many lunar craters. When a massive asteroid slams into the Earth, the target rock behaves like a fluid. It's not unlike dropping a sugar cube into a cup of coffee: the ripples spread outward, but a rebound occurs as the liquid in the center of the cup splashes up. The uplifted rock then solidifies, and the evidence of what happened is preserved. The rim of the whole crater, which was more than 14 miles (22 kilometers) across, has gradually eroded away, although its outline can still be made out when viewed from orbit.

An even larger impact scar, this one in South Australia, was not recognized until the mid-1980s. Maps of that state show many dry salt lakes, one of which looks suspiciously circular. This is Lake Acraman, about 60 miles (100 kilometers) across. All that is left at the impact site is compacted rock from far below the original crater floor, with about a one-mile (1.6-kilometer) depth of rock having been eroded away over many millions of years. But definite proof of the impact remains. A couple of hundred miles away, in mountains known as the Flinders Ranges, rocks have been found that can be linked to the impact site. They are part of the ejecta from the explosion, which flew a huge distance before landing on the ground again. This must have been a truly cataclysmic event.

Into Africa

● **It should be no surprise to hear that this continent too has had its share of major impacts by asteroids and comets.**

Impact craters are well preserved in Australia because of its geological stability and arid climate. Much of Africa is also a useful place to look for the same basic reasons.

Several craters ranging from large to small have been discovered in South Africa. The Pretoria Salt Pan is about 1,200 yards (1,100 meters) across, making it similar in size to Meteor Crater in Arizona. At 200,000 years it is about four times as old as Meteor Crater, and it is more heavily eroded.

Not far away, near the town of Vredefort, southwest of Johannesburg, is a much bigger, older crater. The Vredefort impact occurred almost two billion years ago – before any form of complex life appeared on Earth – leaving a scar almost 90 miles (140 kilometers) across.

Right An aerial view of the Pretoria Salt Pan, a small crater just three-quarters of a mile (1,200 meters) wide.

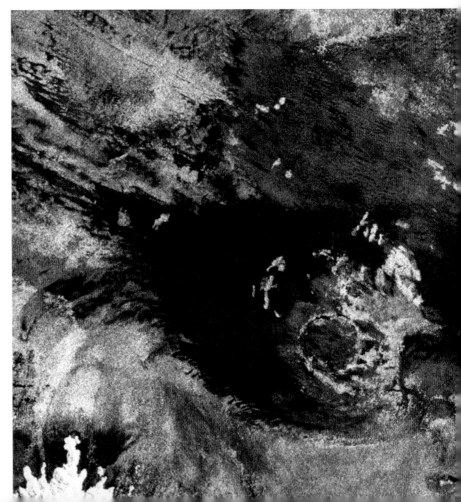

Above The Vredefort impact scar in South Africa photographed from the space shuttle with a hand-held camera. Formed almost two billion years ago, it is about 90 miles (140 kilometers) across.

This is one of the largest known craters on Earth. Had there been animals like dinosaurs alive then, they would have been in big trouble.

The desert in Namibia is host to one of the most spectacular circular impact craters on our planet. Roter Kamm is about 1.5 miles (2.5 kilometers) across, and photographs of it from orbit are extremely reminiscent of the craters of the Moon and Mars that we looked at earlier. Its circular rim is also recognized clearly by satellite-based radar systems, which shows we can apply to the Earth the same techniques that we used to count craters on Venus and elsewhere. A key factor here, and one that makes Roter Kamm so distinct, is that it is relatively young (about five

Above The circular shape of the Aorounga crater is clear in this radar image.

million years old), and is carved out of a region where weathering is slow.

The Deserts of Northern Africa

Skipping over the well-vegetated band of equatorial Africa, the northern region of the continent is covered with desert areas where impact craters may again be expected to show their forms. There are several well-charted craters in North Africa, some of which were found by oil exploration companies.

A recently mapped crater is the Aorounga structure, in an unpopulated and inhospitable part of Chad about 900 miles south of Benghazi, Libya. Formed about 200 million years ago, the Aorounga crater is somewhat more than 10 miles (16 kilometers) across, a conspicuous circle superimposed on the lineations of rock strata stretching across this part of the Sahara.

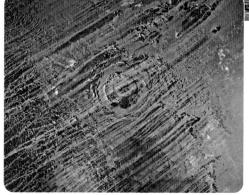

Above A photo of the Aorounga crater in Chad obtained from orbit shows clearly how this crater was blasted out of the target rock.

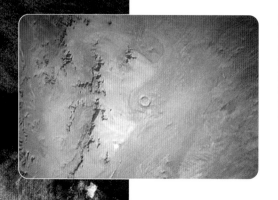

Left The Roter Kamm crater in Namibia, as delineated using a satellite radar (large picture) and also a photograph taken by space shuttle astronauts (smaller image).

Dinosaur Killer

● What happens when a really big asteroid or comet hits the Earth?

The terrestrial craters we have discussed so far have mostly been of modest size compared to some of the behemoths seen elsewhere in the solar system. Once the crater exceeds some tens of miles across, it doesn't make sense to call it a crater. The term used instead is basin.

The lunar maria and Caloris on Mercury are examples of phenomenal impact basins. On Venus, the Mead basin is about 180 miles (290 kilometers) across. Moving outward through the planets, similar huge structures are documented elsewhere by spacecraft imagery. So we need to ask the same question about Earth: where are the vast impact basins you would expect to find?

Extinction Events

This question relates to the mass extinction events evidenced by the paleontological record. Fossil studies show that every so often there has been an abrupt change in the history of life on Earth, some factor that causes what is termed a geological boundary event. At that point in time, the rock strata display a sudden alteration in their nature, providing evidence that the

terrestrial environment changed very significantly during the era in question.

The exact causes of these events in our planet's history have been the subject of argument and debate for many decades. They may look sudden since there is a marked step in the strata. But are they the result of gradual changes, or rapid ones? At one end of the scale, some scientists maintain that the events were slow, with the changes occurring over a few million years as a result of alterations in the Earth's orbit and axial tilt, or due to varying concentrations of oxygen in the atmosphere. At the other extreme are those who claim that the events were very abrupt. In between these two views are those who see the events as rapid on the geological timescale, but slow compared to our familiar time periods. An example is a change in the global environment caused by a period of phenomenal volcanism that lasted for thousands of years – spewing noxious gases into the atmosphere and cooling the climate with the dust the eruptions injected into the stratosphere.

It is widely known that the asteroid or comet hypothesis for the great dinosaur extinction has grown in popularity in recent years. We will look into the history of this concept later. For now, we'll note that the first solid evidence that an impact might have been the cause was found in 1980 in the rock stratum that dates back to 65 million years ago – when the dinosaurs disappeared. And it was discovered at various points around the world. The major evidence was an anomaly in the amount of the element iridium at that precise rock level. Iridium is rare on Earth but common in meteorites. The obvious interpretation was that vaporized

Above Is this how dinosaurs (and pterosaurs) died – in an asteroid-induced cataclysm?
Far left A massive impact elevates millions of tons of the target rock into orbit above the atmosphere, from where it cascades down all over the globe.

iridium from a massive asteroid strike had settled all over the planet. Calculations show that the impactor must have been between 5 and 10 miles (8 and 16 kilometers) wide, producing a crater 100 miles (160 kilometers) or more in diameter. But where was it? The search was on.

The Smoking Gun

Several craters were put forward as possible sites for the impact that killed the dinosaurs. The Manicouagan crater in Canada and Popigai in Siberia were well known and, at about 60 miles (100 kilometers) wide, almost the right size. But one is too old, and the other is too young. The Manson crater in Iowa seemed to be about the right age, but at only 22 miles (35 kilometers) across, it was too small. More recent dating has shown that it was formed about ten million years before the mass extinction event.

It was in 1990 that the location of the dinosaur-destroying impact became widely known. In fact, the existence of an anomalous zone at the north end of Mexico's Yucatán peninsula had been common knowledge in certain circles (such as the oil exploration industry) for some time, and was suspected of being evidence of a phenomenal extraterrestrial impact. It took a while, though, for the anomaly to be connected with the dinosaur extinction. This structure is called the Chicxulub basin. It is the smoking gun that points to how the dinosaurs died.

Right These glassy granules collected in Mexico are melted target rocks thrown out by the Chicxulub impact.
Bottom right This tiny quartz grain, only one eightieth of an inch (one-third of a millimeter) in size, was drilled out from the Chicxulub impact site. It shows clear evidence of massive shock compression, stressing its crystal lattice.

Left By plotting the gravity field in the region, geologists can get a good idea of the extent of the Chicxulub crater, as these maps show.

Rain of Fiery Rock

The explosive energy of the dinosaur-killing impact was sufficient to hurl rocks all around the globe.

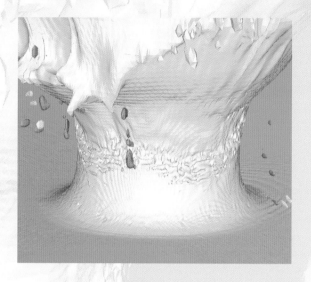

The size of the Chicxulub basin is the subject of continuing debate among impact geologists. Some think it is only 120 miles (200 kilometers) across, and that arcuate or curved cracks in the rock strata a good distance away are faults formed by the shock waves generated in the impact. Others claim that the original crater cavity was actually around 200 miles (320 kilometers) across, and the cracks represent the real rim before infall filled this vast hole.

Either way, this was clearly a gigantic impact. Best estimates calculate the amount of energy released at around the equivalent of one billion tons of TNT, which corresponds with the impact of a rocky object about 10 miles (16 kilometers) in diameter. There is some evidence that the impactor might have been not an asteroid, but a comet. That would increase the likely impact speed, meaning that an object of lesser mass would be required to release the same energy, but the lower density of a comet – composed largely of ice – means that the estimated size remains at about 10 miles.

The initial excavated crater would not only have been broad. It would also have been deep – a dozen miles or more of the crustal rock

Right, and background Scientists now have the ability to simulate with computers the effects of massive impacts on land or in the ocean. **Sequence below** An impacting asteroid throws out a huge mass of rock. A thousand years later, all that is left is a huge crater. Much later still, even the crater is difficult to recognize.

would have been instantaneously broken up and hurled outward. As much as 100,000 cubic miles of Mexico may have been pulverized and ejected.

The explosion would have blown a hemispherical hole in the atmosphere through which many of these stone fragments would have flown. Some would have had speeds in excess of the escape velocity, and these would have hurtled off into interplanetary space, perhaps to arrive on another planet one day. But much of the ejecta from the crater would be traveling less rapidly, and would have rained back on Earth, landing in every corner of the globe.

Within a few minutes, rocks would have been cascading down on the far side of the Gulf of Mexico. Within an hour, similar lumps would

have reentered the atmosphere above India, Southeast Asia, and Australia. There was nowhere to escape. Nothing was safe.

Global Inferno

How did the dinosaurs die? In simple terms, they were grilled to death. This volume of ejecta cascading down into the atmosphere at typical speeds of 5 miles (8 kilometers) per second possessed huge amounts of energy. Like meteoroids from space producing shooting stars, the reentering rocks ablated or burned up in the atmosphere at altitudes between 60 and 20 miles (100 and 30 kilometers). Some large rocks may have penetrated deeper, giving rise to blasts like the one over Siberia in 1908, but the major killing mechanism originated from the trillions of smaller lumps ablating higher up.

These dissipated their energy through the emission of visible and infrared radiation. It was like a meteor shower in which the whole sky was covered; a gigantic grill stretching from horizon to horizon, far brighter than the Sun. The temperature on the Earth's surface would soon have reached a thousand degrees.

Animals in the open were fried. The only species that seem to have escaped were those that weighed less than 50 pounds (25 kilograms) – some of them may have survived in burrows, holes, or tunnels – or those that lived deep in the sea. There were few sanctuaries.

Plant life rapidly desiccated, even in dense tropical jungles, and burst into flames. The geological stratum representing the end of the dinosaur era contains a layer of soot and charcoal that points to the incineration of more than 90 percent of the planet's biological mass. Many animals that survived the initial onslaught would subsequently have starved to death.

But their problems didn't end there. For a few days or weeks the rocks rained from the sky and the flames burned on the ground, before

the situation cooled off. But it cooled off too much. Dust of pulverized rock filled the upper atmosphere, and was joined lower down by soot from the global firestorm. The level of sunlight reaching the planet's surface dropped precipitously, and global temperatures plummeted.

The Cosmic Winter

It takes many years for so much dust to settle. By that time ice and snow would have covered much of the land and oceans. Because snow reflects most of the sunlight that hits it, the low temperatures were maintained, and the Earth entered an Ice Age. Those species of flora and fauna that had managed to survive the heat now needed to battle the cold.

There were other upsets too numerous to describe in detail. For example, the target area happened to be rich in sulfate rocks, and the impact released sulfur-bearing gases into the atmosphere. These combined with water to produce noxious chemicals including sulfuric acid, which then fell to ground as highly concentrated acid rain.

No wonder so few species survived. But as one order of life passed away, so niches opened up for others to fill. This led to the rise of the mammals. And, in time, to us.

Background Global fires would be triggered by a massive impact. Geological investigations have shown that this happened at the time the dinosaurs died. **Above** Subsequent to an impact, scientists expect the climate to cool, producing a "cosmic winter."

Impacts and Tektites

Tektites are strange, abnormal pieces of rock, the result of massive impacts. Smooth and glassy, they range from microscopic to the size of an egg. The larger ones come in a variety of shapes including spheres, dumbbells, and buttons.

Background A variety of aerodynamic shapes of tektites collected from sites in Texas, Vietnam, and Australia, ranging up to an inch or so in size. The semitransparent sample on the right is an example of the so-called Darwin Glass found in Tasmania, apparently from a different impact to the one that showered much of southeast Asia and Australia with tektites.

Tektites are found scattered over many thousands of square miles in distinct zones spread around the world. Each set of tektites – called a strewn field – can be dated as having been formed at a particular time, which shows that an infrequent event was responsible in each case. Millions of years separate the ages of these strewn fields.

Tektite Origins

Their origin has been a long-term puzzle for geologists because they show no obvious relationship to the native stone wherever they are found. They are dark, and usually gray-black in appearance, although a greenish tinge is sometimes reported. This makes them obvious in light, sandy soil. Their glassy nature is a clear sign that the constituent material had melted, and their aerodynamic shape indicates that they cooled and solidified from a molten state while flying through the atmosphere.

The wide scatter of tektites shows that whatever liquefied them and threw them outward must have been a highly energetic explosion. An asteroid or comet impact is an obvious candidate. But if the atmosphere had been dense at the start of their trajectory, then they would have been torn apart into tiny droplets with no large lumps remaining. This suggests they must have cooled while entering the atmosphere from above.

How could that be? One hypothesis says that they are lunar fragments that were sent earthward following an impact on the Moon, but there are several objections to this idea. If this were the case, then you would expect the tektites to be spread all over the planet and not restricted to certain locations. Also, they would have cooled and solidified in space, and

we would expect them to be spheres rather than the aerodynamic shapes actually identified.

The answer seems to be that tektites are the fingerprints left by very large impacts on the Earth. If a projectile were massive enough – a few miles wide – then it would blow a cap off the atmosphere above the impact site. On the way down, it would also push an empty column through the air, and it would take about ten seconds to refill those empty volumes. During that brief time, the projectile would have hit the target rock, excavated a crater, and ejected globules of molten glass outward above the atmosphere. On their way back down, the globules would cool and solidify, eventually landing over a wide area, leaving enigmatic clues about their origin.

All Over Dixie

One of the best-known tektite regions is the North American strewn field, which has been known for decades. From Virginia down through the Carolinas and Georgia, and all across the southeastern United States to Texas – in fact, all over Dixie – you can find tektites in vast numbers if you look in the right places.

Many of these have a common age of about thirty-five million years. This is also the age of the Chesapeake Bay impact structure. It seems evident that the impact left not just a crater but a vast array of debris spread over an extremely wide area – unambiguous evidence that the explosion was phenomenally energetic.

A bit farther south is another strewn field, concentrated around Haiti, although samples have also been collected from other Caribbean islands, as well as the Mexican coast. These have a common age of sixty-five million years. They are clearly pieces of the Yucatán peninsula that hurtled skyward when the dinosaur killer struck.

Europe has not escaped. We read earlier about the Ries basin, a scar formed by an impact in Germany around fifteen million years ago. This seems to have produced a group of tektites called the Moldavites. In Belgium the buried strata from 365 million years ago bear a concentration of tiny microtektites, which seem to be linked to the impact that produced the 34-mile (55-kilometer) Siljan Ring in Sweden.

The best terrain for finding tektites is hard, dry ground, preferably light in coloration, and unfettered with vegetation. That's a description of much of Australia. Much of that island continent is strewn with tektites, apparently produced by an impact 700,000 years ago in Cambodia. The impact left an elongated scar that is today a lake called Tonle Sap.

Above Almost spherical microtektites from the dinosaur-killing impact in Mexico, collected in the rock strata of Wyoming, 1,800 miles (2,900 kilometers) from the crater. These are about forty thousandths of an inch (one millimeter) in size.

Mega-Tsunami

● **An oceanic impact would produce a huge tsunami that would sweep unchecked around the globe.**

Tsunamis are often called tidal waves, although they have nothing to do with the tides. The daily ebbs and flows of the tides are due to the gravitational pulls of the Moon and the Sun, which are smooth, steady, and predictable.

Tsunamis are quite the opposite: they are unpredictable and devastatingly violent. Generally caused by earthquakes or suboceanic landslips, tsunamis occur around the world on a yearly basis. A couple of years back a tsunami struck the north coast of New Guinea, killing thousands of people living by the sea.

The Pacific is a particularly dangerous place to live in this respect, because a tsunami can sweep across its width unchecked. In 1960, a small earthquake off the coast of Chile caused a wave that killed at least a thousand people on the nearby mainland. The surge then traveled across the Pacific, reaching Hawaii fourteen hours later. In the open ocean it was a comparatively low profile wave – a swell just 15 centimeters (6 inches) or so in magnitude. As it hit the coastal shelf off the islands it ramped up to a height of 30 or 40 feet (9 to 12 meters), and killed sixty-one people in Hilo harbor. When it reached Japan, the wave drowned another two hundred people.

Similarly calamitous events have happened in Alaska because of seismic activity in the Aleutian Islands. Tsunami warnings are a fact of life for coastal communities in parts of the northern Pacific. But even practice drills could not protect them against an impact-induced tsunami.

Impact Tidal Waves

That impacts could cause huge numbers of deaths through tsunamis was not recognized until just recently. There seems to be a widespread perception among the general public

Above Boats were beached and trucks and buildings wrecked by a tsunami in Alaska in 1964.
Right Damage caused in Japan by the 1960 tsunami, which was generated on the opposite side of the Pacific by an earthquake near Chile.

that an impact in the ocean would pose less of a hazard than an impact on dry land. Astronomers counter that oceanic impacts are every bit as dangerous. Someone calculated what would happen if a comparatively small asteroid were to plop into the Pacific. The answer makes alarming reading for anyone living anywhere around the Pacific Rim.

There is evidence that such occurrences have happened in the past. In Hawaii coral has been found 1,000 feet (300 meters) above sea level. On the eastern coast of Australia house-sized boulders apparently plucked from the sea floor have been found on top of cliffs 40 feet (12 meters) high. In northwest Australia, there are signs of a tsunami that swept 20 miles (30 kilometers) inland from the coast, supposedly when an impact occurred in the Indian Ocean. And these must have been in fairly recent times.

If you live in San Francisco, a major earthquake is not the greatest natural hazard you face, since a phenomenal tsunami generated by a comparatively small impact may occur at any time. But residents of Boston should not feel too smug. The Atlantic is also a vast target. A 200-yard (180-meter) asteroid slamming into that ocean would wipe out every city on the eastern seaboard of the United States. Lisbon in Portugal and Casablanca in Morocco would follow, and as the wave swept up the coast of Europe it would cause devastation for tens of miles inland over the coastal plains. You can bet that the dikes of Holland wouldn't hold it back.

The projectile doesn't even need to reach sea level to generate a tsunami. The Tunguska event in 1908 (see page 86) led to a shock wave that flattened the forests of Siberia. If that boulder had arrived a few minutes earlier it would have been over the northern Pacific. Instead of blasting trees down, its shock front would have punched into the sea, causing an effect like that produced when you kick a bucket of water.

Within hours, a tsunami unlike any in human memory would have inundated the coasts, from Japan and Korea to Vancouver and San Diego.

We were lucky in 1908. About the worst place for an impact of any size is the open ocean. The sea is deadly efficient at transporting energy from one place to another. And I do mean deadly.

Impact Tsunami

As these computations show, the waves produced by an asteroid hitting the ocean are similar to those formed by throwing a stone into a pond, but on a vastly larger scale.

Initially the ripples, which may be hundreds of yards or meters high, spread out as regular circles (top diagram), but gradually they distort as they meet land or the varying depths of the sea, affecting the speed at which the wave moves.

A suboceanic crater recently found under the southeastern Pacific indicates an impact by a 2.5-mile (4-kilometer) asteroid about two million years ago. Computations of the tsunami it must have produced (middle diagram) indicate that, five hours after the impact, a wave 60 to 80 yards or meters high was spreading across the Pacific and had also penetrated into the South Atlantic.

As time progressed, it reached coastlines spread from Australia to Africa in the southern hemisphere, and after twenty hours it swamped the northern Pacific rim from Japan to Alaska.

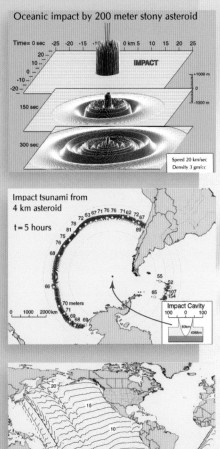

Oceanic impact by 200 meter stony asteroid

Speed 20 km/sec
Density 3 gm/cc

Impact tsunami from 4 km asteroid
t = 5 hours

Impact Cavity

Driving Evolution

● **Catastrophic impact is now being recognized as one of the controlling factors in the evolution of life.**

Competition between producers drives a market economy. This is known as Adam Smith's invisible hand. Similarly, according to Darwinian theory, competition between slight genetic variants drives evolution. But was Charles Darwin entirely correct?

From the scientific perspective it seems that Darwin was generally right, but incorrect in some of the details. In the mid-nineteenth century Darwin, influenced by his geologist friend Charles Lyell, thought that evolutionary change plodded along slowly and steadily, with little or no apparent change from one

generation to the next. New species gradually came about over long periods of time. To Lyell, geological change was brought about by the accumulation over long periods of the everyday effects of the agents of change. Thus a river flowing for millennia would delineate a valley through the systematic erosion of soil and sand.

But Lyell was wrong. The major contributor to valley formation is not the everyday flow but the occasional flood of the river – the once-a-century or once-a-millennium tumult that dominates the net rate at which solid material is washed away. Similarly, evolution occurs not at a

Below The tree of life for the past several hundred million years. Evidence is growing that the abrupt extinctions of some family groups, followed by the blossoming of others, are generally linked to massive impact events.

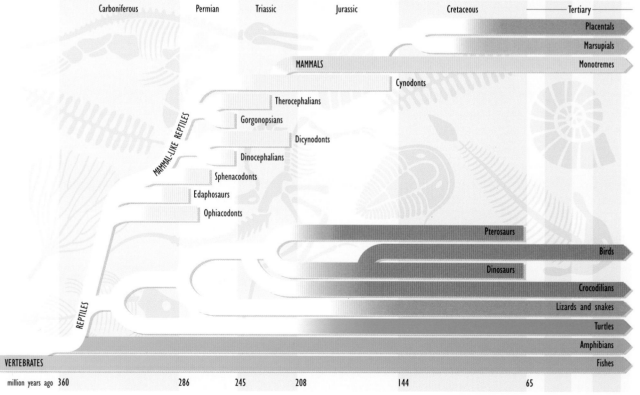

steady rate, but as a series of stops and starts. This is called punctuated equilibrium. Nothing much happens until there is a drastic change.

Colossal Upsets

When it comes to evolutionary change, asteroid collisions represent colossal upsets. All of a sudden, the basic rules of survival change. Animals are exposed to an environment that is entirely unfamiliar, a set of conditions for which natural selection has not occurred. Nothing can prepare a species for the cataclysm of an asteroid impact.

Take mammals, for example. During the dinosaur era they seem to have been small, and thus able to hide from larger predators within burrows, just as rabbits escape foxes today. Although this selection for diminutive size was based on avoiding predators, it served them well when a huge asteroid struck the Earth, burning everything on the surface. Certainly the vast majority of individual mammals and most mammalian species perished, but it seems a small fraction survived the initial onslaught underground. Then, thanks to their furry coats, their omnivorous nature, and their ability to regulate their temperature, some eked out a living during the subsequent cosmic winter.

But they had not evolved in order to survive an impact. That was coincidental.

Bad Genes or Bad Luck?

How does one view this phenomenon in terms of evolution? David Raup, a prominent paleontologist, posed the question by asking whether much of evolution is a case of bad genes or bad luck. That is, did the dinosaurs die out because they were inadequate and so outdone by some other family of beasts? Or were they simply unlucky?

Since the dinosaurs were the dominant group for many millions of years – far longer than

humanity has walked the Earth with its bipedal gait – you could say that the terrible lizards' luck eventually ran out. But there is a moral here for us, if we care to see it.

A major problem in evolutionary studies is how mankind developed mathematical abilities. It does not seem to give us a survival advantage in the same way as sharp claws and teeth as weapons, or a shell for protection, aid the reproductive success of an animal. How and why did we develop the ability to do mathematics and consequently science that have together enabled us to move so far and so fast?

The fact is that we do have the set of genes required for such reasoning, and because of that we have the ability to circumvent the dilemma of bad genes or bad luck. We have the genes that enable us to do what the dinosaurs could not. We can see our asteroidal nemesis coming and we can do something about it.

Life has existed on our planet for about 3.8 billion years. Until 570 million years ago it was just slime, before a sudden blossoming of life occurred in the sea and the first polycellular creatures appeared. Eventually, complex life migrated to dry land. Its development underwent many setbacks, as impacts and other events shifted the environmental rules. Each change created empty niches in which new species could evolve. The dinosaurs would never have entered the picture if the previous top predators had not been extinguished by an impact about 215 million years ago.

Eventually a species arose with the capability to take its destiny into its own hands, and ensure that its days are not brought to a premature end by some rogue asteroid or doomsday comet. We are that species.

Background The fragile Earth. The heavily-cratered Moon tells us that cosmic impacts are a fact of nature, and now we are finding more and more evidence of such impacts on Earth. Do we live on a safe planet?

Rain from Space

The solar system is teeming with lumps of cosmic debris that can and do slam into the planets. Some of these objects are large, punching through atmospheres to produce cataclysmic explosions that leave huge craters. Some are small, perhaps the size of a grape, and these completely burn up high in the firmament, their death throes lighting up the night sky as shooting stars.

Others are smaller still, and the ends of their interplanetary trajectories are anything but violent. They lose their energy gradually in the upper reaches of the atmosphere, decelerating from their phenomenal cosmic speeds until they come to a stop, before settling out to the ground. For this to happen, a particle needs to be smaller than about a thousandth of an inch (a few tens of microns) across so that it behaves like a mini-parachute. Any larger and the friction with the air on entry will first melt it, then make it evaporate.

To reach the ground intact, an impinging body must either be big – in which case it is instantly destroyed in its hypervelocity impact with the solid Earth – or very small, which makes a slow descent possible.

Earth Putting on Weight

How much solid matter does our planet accumulate in this way? The answer depends on the size of the objects of interest to you. The total influx of small particles – meteoroids and interplanetary dust ranging from the size of a basketball down to the tiniest grains – amounts to about 40,000 tons each year. That's a very considerable mass, made up of many trillions of these tiny bodies. In terms of massive asteroids and comets, the incoming mass in most years amounts to zero. But every so often we get a big lump. In the long run, these larger objects dominate the mass influx.

Although this implies that the Earth is putting on weight at a considerable rate, we should note that there are also ways in which its mass drops. Very light gases, like helium, for example, gradually permeate to the top of the atmosphere and then leak away into space.

The Hard Rain

Much of the annual 40,000 tons of influx burns up in the atmosphere, providing free sky shows of the type described on the next few pages. But a small fraction slows down and reaches the ground more or less intact.

These particles accumulate in various places on the Earth's surface. Over a century ago one of the first great oceanographic expeditions, the voyage of a British vessel called the *Challenger*, led to the discovery of debris from outer space at the bottom of the ocean. The Victorian scientists trailed a large magnet on a rope across the floor of the Pacific. When they brought it to the surface they found it covered in myriad tiny globules of ferric metal from the muddy bed of the sea. These were spherical in shape and correctly identified as melted micrometeorites.

Such metallic grains, though, comprise only a minor fraction of the total influx of tiny particles. Most are stony in composition, and so not attracted to a magnet. These can be isolated in other ways: for example, by filtering them out of Antarctic ice as it is melted. Over many millennia significant quantities of such dust from space have accumulated in the polar ice caps.

To get really pristine samples, the micrometeorites must be collected before they reach the ground. This is accomplished using sticky-plate detectors slotted into the wings of very high-flying aircraft. Once the planes reach an altitude of about 10 miles (16 kilometers) – above the dust whipped up from the ground, and the exhausts of commercial jetliners – these plates are exposed to the airflow, and then retracted before descent. Under the microscope they reveal many particles that are clearly of extraterrestrial origin.

Left A true-color image of Comet Halley built up from three separate red, green, and blue photographs. Such comets produce most of the meteoroids and dust in space. The colored streaks are background stars.

Right Even dust grains can damage satellites, due to their high impact speeds. This photograph shows the tiny crater punched into the solar cells of the Hubble Space Telescope by such a dust particle.

Above A tiny interplanetary dust grain, photographed using an electron microscope. This particle was collected in the upper atmosphere using a NASA sampling aircraft.

Shooting Stars

A piece of cosmic rock the size of a pea entering the atmosphere possesses as much kinetic energy – the energy of motion – as a stick of dynamite.

At a typical altitude of 80 miles (130 kilometers), the frictional drag of the air begins to heat it up. At 60 miles (97 kilometers) it is glowing, and molecules are boiling off from its surface. In a blaze of glory we see it zip across the starry sky: by the time it reaches an altitude of 40 miles (65 kilometers) it is all over. The pea-size rock has disappeared, its constituent materials dispersed forever in the upper atmosphere.

Such an object is called a meteoroid while it is in space; the phenomenon we see as it dies is called a meteor.

A meteor is not a solid body: the word correctly applies to what is seen by eye, or detected using other instruments such as cameras, video systems, or radars. A more common term is a shooting star.

Random Meteors

At a good, dark site, on any night, an individual with average eyesight should see about ten shooting stars per hour. The best time to look is just before sunrise. The influx of meteors in the atmosphere above you peaks when you are at the leading point on the Earth in its orbit, which is the 6 A.M. position. Unfortunately, many people tend to look for meteors soon after dusk, which is just about the worst time: at 6 P.M. you are on the trailing side of the planet, and much less likely to witness the sight.

Above A long-exposure celestial photograph smears the stars out into curved lines. Very briefly, shooting stars zip across the sky, producing long straight images. These are dashed because of a rotating shutter in front of the camera lens, making it possible to measure their speeds.

This variation is analogous to the pattern of rainfall on a car's windows. As you drive along, more raindrops hit the front windshield than the back. Radar systems can be set up to patrol for meteors twenty-four hours a day, unaffected by daylight or cloud cover. The data from these systems confirm that the observed frequency is more than merely a quirk of visibility: only about a third as many meteors are detected at 6 P.M. as at 6 A.M. For the best chance of seeing shooting stars, you need to be out and about in the early hours.

These meteors appear in random parts of the sky, traveling in various directions. Astronomers call them sporadic meteors. In essence, the meteoroids are tiny fragments of long-dead comets, or chunks lost by asteroids. Over millennia, their orbits have been stirred up by the planets and buffeted by collisions with dust in space; their arrivals are uncorrelated.

Cometary Fragments

Other meteoroids are not random at all, but move in complexes of common origin.

Imagine the dust tail of a comet. This is made up of minute fragments expelled by the expanding gas from the cometary nucleus, which is like a dirty snowball. It is a solid mass of ice and other frozen components, with dirt and rocky debris thrown in. Although most of the solid pieces are dust grains that are soon dispersed by forces such as the light pressure imposed by solar radiation, some are larger and can measure up to an inch (2.5 centimeters) in size. These tend to follow the same orbit as the comet, although the low speeds at which they are ejected means that after a few centuries they begin to spread out at staggered intervals behind the comet on its orbit of the sun.

Over time the trailing meteoroids stretch out to form a complete loop along the comet's orbital path. Although spread across a vast segment of the solar system, they form a coherent structure because they are all moving along basically the same path. When the Earth traverses one of these loops, many meteoroids enter the atmosphere moving with parallel paths and traveling at the same speed. We call this spectacular sight a meteor shower.

Meteor Showers

Most of the pieces of cometary detritus that produce a meteor shower are smaller than a peanut. Despite this, they produce a vivid effect as they zip into oblivion. Because their paths in

Right Comets have two distinct tails, as seen here. The ion tail, made up of charged atoms, appears blue and stretches away from the direction of the Sun. The dust tail, containing the small solid lumps that eventually produce shooting stars visible on Earth, is slightly pink in color and has a fan-shaped profile lagging behind the comet. The red smudge at far left is a distant galaxy.

Below A bright meteor speeds across the pre-dawn sky, announcing the death of a tiny rock from space.

space are parallel, they seem to emanate from the same point in the sky, known as the shower radiant. The phenomenon is similar to viewing a set of railroad tracks. You know the tracks are parallel, but they appear to meet in the distance.

Meteor showers are named after the constellation that contains their radiant. So, for example, the Perseid meteor shower, which occurs every year with peak activity around August 12, has its radiant in the constellation Perseus. This does not mean that the meteoroids originated there: it is purely a matter of perspective. The Perseid shower, we know, is made up of fragments of Comet Swift-Tuttle.

Other well-known showers include the Geminids (peak near December 13), the Quadrantids (January 4), the Lyrids (April 22), and the Taurids (activity peaking at the end of October or beginning of November). Comet Halley produces two meteor showers every year: the Eta Aquarids in the first week of May, and the Orionids in the third week of October. Typical count rates during a meteor shower are twenty to fifty per hour.

When is the Earth due to be hit by a comet? The answer is that we are struck by bits of comet all the time. Similarly, you don't need to wait until 2062 to see Comet Halley again: you can see pieces of it when its meteor showers occur twice a year.

Meteor Storms

Very infrequently, the sky lights up with a supercharged meteor shower.

The annual meteor showers we've discussed show much the same level of activity from one year to the next. Slight variations do occur, however, due to factors like the brightness of the Moon, which can mean that the very faintest shooting stars are not detectible with the naked eye. These regular showers, as we have seen, are caused by a trail of tiny lumps shed by the parent object spreading out along the comet's orbit in a complete loop. We know it has been some millennia since Comet Halley released the particles that are capable of hitting us now.

Close Encounters

This is not the case, though, if a comet has a trajectory leading it to cross the ecliptic close to 1 AU. In that case, even recently released meteoroids can intercept us.

A good example is Comet Swift-Tuttle, the parent of the Perseid meteor shower. Observations made in the 1860s, on the last occasion it passed by, indicated that this comet would return in 1980. As we now know, it did not appear on schedule. But in the early 1970s, Brian Marsden, who contributed to the foreword of this book, computed other possible orbits for the comet. He suggested that if it failed to appear in 1980, then it would be seen in 1992. Events in the late 1980s lent weight to his theory: the Perseid meteor shower activity was noted to be enhanced, suggesting that the comet was indeed on its way. What was

happening was this: the comet's daughter meteoroids, released on its last flyby in the 1860s, were dropped into slightly smaller orbits, which meant that they were now preceding their parent, thus heralding its approach. Sure enough, Swift-Tuttle was soon spotted, exactly as Marsden had predicted, and throughout the 1990s the Perseids were more prominent than they had been a decade or more earlier. This is because there is a concentration of solid debris relatively close to the comet that has yet to be dispersed around its orbital loop.

Data gathered through observations of Comet Swift-Tuttle in 1992 have made it possible to link it to a poorly observed comet in 1737. Chinese records dating back to 68 B.C. also provide evidence of links with the comet.

Above This photograph taken through a fish-eye lens shows the Leonid meteor shower, with shooting stars appearing to emanate from its radiant in the constellation Leo.

The Lion Roars

When the Earth happens to pass very close to a comet nucleus, the associated meteor shower is often amplified, producing what is known as a meteor storm. Precisely what counts as a storm is rather subjective, but meteor watchers use the term to describe a shower with in excess of a thousand shooting stars per hour. The Perseid meteors produced strong showers of a few hundred per hour, but no real storm.

Another comet codiscovered by Horace Tuttle does produce storms, though. This is Comet Tempel-Tuttle. Nowadays, this comet is responsible for an annual meteor shower that occurs around November 17 or 18, called the Leonids, emanating from the constellation Leo, the Lion. Every thirty-three years – the orbital period of the parent comet – the lion roars and a meteor storm is seen.

Such events are recorded as far back as A.D. 902, but the best-known showers have occurred in the past few centuries. In 1799, a mighty display was seen in the eastern Americas, and again in 1833. In 1866, Europe was the place to be. After disappointments in 1899, 1900, and 1933, the Pacific coast of North America was treated to a phenomenal storm in 1966.

There are two points to note from these records. First, the cycle is not exactly thirty-three years. Nature is more complicated than that. Second, because it takes the Earth only a few hours to travel through the densest concentration of meteoroids in space, you have to be at the right place at the right time to see the real storms.

In 1999, the prediction was the storms would be visible for longitudes from the Middle East throughout Europe. If you were farther west, the Moon would interfere. Farther to the east and Leo would have set already. Sky watchers who went to the predicted best viewing locations were not disappointed: a great number of shooting stars lit up the heavens bang on schedule.

Watch This Space

The Leonid meteors in 1999 represented something of a triumph for theoretical astronomy. We knew that a storm was due, but beyond this, any more specific predictions had

Below and background An impression of how the sky appeared over Pennsylvania in 1833, when the Leonid meteor storm provoked a huge commotion.

always been largely a matter of guesswork. This time, though, my colleagues David Asher and Rob McNaught had constructed a computer model showing how meteoroids were thrown out by comets, enabling them to follow their paths forward over some centuries. The model was detailed enough for them to determine that the particles hitting the Earth in 1999 were liberated by the comet three orbits before. They also calculated to within a matter of minutes when the peak activity would occur. And they got it absolutely right.

The accuracy of their model gives us the confidence to predict how the Leonids will behave over the next little while. It seems likely that there will be vivid displays in 2000, 2001, and 2002, and then a pause until 2006. Such a gap is not unexpected: the storm of 1966 was followed by an outburst in 1969. So it is worth planning to go sky watching in mid-November in the coming few years.

The Leonids are not the only meteor storms to be seen. Several other comets, including Comet Giacobini-Zinner, produce spasmodic storms, but none is expected soon. On the other hand, history shows that every decade or so we experience a meteor outburst of unknown origin, indicating that there are massive parent objects flying close by that we have yet to spot. We ought to regard these meteors as shots across our bows, warning us of potential trouble.

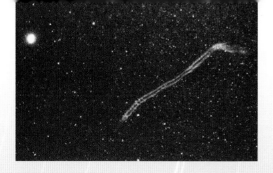

Fireballs and Flashes

● **The brightness of a shooting star depends upon such factors as the size, composition, and speed of the incoming particle.**

Above This bright fireball "spoiled" a deep sky photograph taken with the U.K. Schmidt Telescope in Australia, appearing to bisect the galaxy at top left.
Above right Very bright meteors can leave glowing trails in the atmosphere, which persist for many minutes as winds distort their shape.
Background A fireball that exploded in a terminal flare. Taken in the Czech Republic, this long-exposure photograph records stars as long arcuate trails, due to the Earth's rotation.

Consider a marble-sized meteoroid entering the upper atmosphere at 12 miles (19 kilometers) per second, a typical arrival velocity. By virtue of its high speed, it possesses the same kinetic energy as a bowling ball traveling at more than 3,000 miles (5,000 kilometers) per hour. That's a lot of energy to be dissipated within a second or two in the upper atmosphere. No wonder the meteoroid gets so hot that it melts and evaporates, ending its life as a glowing streak 40 miles (65 kilometers) or so above our heads.

Typical naked-eye meteors result from original particles smaller than a pea. That marble-sized visitor would be fairly bright if you were out scanning the sky on a dark night. But larger meteoroids – perhaps as big as an orange or a grapefruit – would catch your attention even if you were not specifically looking for them. These bright meteors are called fireballs, or bolides.

As Bright as the Sun

Fireballs often startle people who are driving at night in unpopulated areas, or camping in the countryside where there are no houses or streetlights to diminish the darkness. Many of us live in towns or cities surrounded by artificial lighting, and over the past century we have lost our familiarity with the sky. We are then surprised to see celestial phenomena that would have been commonplace to our ancestors.

Back in April 1993 there was a particularly bright fireball over southeastern Australia. I noticed it light up the night sky while I was watching TV. The flash it produced was sufficient to cast shadows inside the house for a brief

instant. From good eyewitness reports, we were able to determine that it had finally expired a dozen miles above the gold mining town of Peak Hill, about 200 miles (300 kilometers) north of Sydney. Many witnesses were fooled, though, thinking that the fireball was much closer to them than it really was. Because people are used to seeing high-flying jet aircraft moving across the sky at a certain speed, they interpret a fireball's rapid motion as evidence that it is closer. In fact, a fireball is typically seen ten times higher than a jumbo jet, and it travels a hundred times faster.

The energy posessed by these lumps is huge, and a basketball-size meteoroid may light up the evening darkness so that it looks like daylight for a fleeting instant. They can be as bright as the Sun. Although the flash doesn't last for long, a fireball can leave behind a trail that will continue to glow for minutes, or even hours.

Looking Down

Fireballs are seen every day in many places around the globe, but most are missed because they enter the atmosphere over sparsely populated regions. Cloud cover also limits their detection: meteoroid ablation is finished while the object is still far above the cloud deck. In addition, the meteoroid influx peaks close to 6 A.M. An equal number arrive during the daytime, when they are largely invisible. Although a very bright fireball may be seen on a clear day, I have witnessed only one in my entire life.

Such sightings involve our looking up into the

sky. Since the start of the space age we also have had satellites in space looking down, scanning the globe for various purposes, such as land resource surveys and weather prediction. Bright meteors must be detected every so often by these satellites. Even more significant are surveillance systems operated by the militaries of space-faring nations – particularly the United States – which depend on satellites for rapid and frequent monitoring of the whole globe.

As far as the detection of fireballs is concerned, "looking down" from orbit has several advantages over "looking up" from the Earth's surface. Cloud cover is not an issue, there's no rain or wind to worry about, and there are other benefits. The radiation emitted by a fireball may be largely in the ultraviolet and infrared regions of the spectrum. Such UV and IR traveling downward are absorbed by the atmosphere and are not detected. The upward emission passes through the thin upper atmosphere, where it is largely unimpeded and can be detected by satellite instruments in orbit.

False Alarms

Fireballs appear in satellite cameras as sudden flashes characteristic of explosions of material at temperatures of a few thousand degrees – just like the rocket launches that many surveillance satellites are designed to monitor. Imagine that, in a time of tension, country X has satellites looking down on country Y, because it fears a pre-emptive nuclear attack. Those satellites detect a flash in the region of one of the launch bases of country Y. How will country X respond unless it is aware that the flash may be a perfectly natural occurrence? Or, what if a meteoroid delivering the energy equivalent of the Hiroshima bomb were to explode above the capital city of nation Z, making the government think it's under attack? Clearly we have many reasons for needing to understand these flashes in the sky, not the least of which is to prevent the outbreak of World War III due to a tragic error of interpretation.

Below A map of bright fireball flashes recorded by U.S. surveillance satellites over the past two decades.
Bottom A lump of rock from space breaks up as it plummets through the atmosphere. This multiple fireball was seen over a wide area of the eastern United States in 1992. One fist-size fragment fell in Peekskill, New York, punching a hole through the trunk of a car.

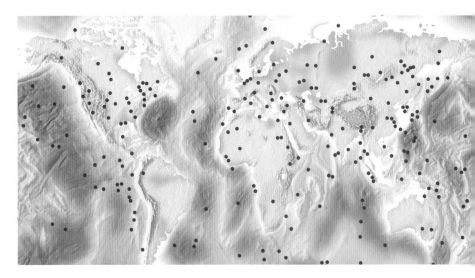

Puncturing Satellites

Space is not a safe place. A grain of interplanetary dust could kill an astronaut on a space walk, or put a satellite out of action.

Looking down from orbit, satellite surveillance systems pick up many fireballs. Large meteoroids entering the atmosphere at high speed produce these flashes. But there are many more small particles arriving, most of them too faint for us to see, and these might hit a satellite.

The tiniest grains slow down in the atmosphere without being melted, and they gradually settle to the surface. Researchers may then collect them; for example, they may melt large amounts of polar ice and sieve the micrometeorites out from the water. This does not tell us when the particles arrived, though, nor their speeds and so on.

One way of detecting interplanetary dust as it nears Earth is to put a suitable instrument on board an orbiting satellite. When the first such experiments were done in the 1960s, the detectors were quite simple: basically small microphones that would pick up the "pings" that resulted when particles traveling through space struck the surface of the satellite. Later, more sophisticated systems were flown, such as electronic devices measuring the ionization produced whenever an interplanetary dust grain zapped into them.

Below left A microcrater punched into part of the Hubble Space Telescope, returned to the ground for study after a servicing mission.
Below A crater in an aluminum surface from LDEF, produced by a dust grain striking it at high speed.

Left High above Baja California, the robot arm on the space shuttle grapples with the LDEF (Long Duration Exposure Facility) satellite. LDEF, which was left in orbit for six years, was half the size of a bus, and returned peppered with craters and punctured metal surfaces, telling us much about the flux of small solid particles in space.

Below The material ejected from a tiny impact crater was splattered over an adjacent surface when a meteoroid slammed into a corner strut on LDEF.

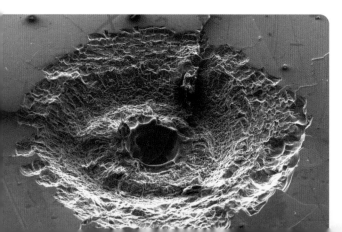

Recovered Hardware

Obviously there are reasons to be interested in the flux of interplanetary dust and small meteoroids apart from abstract scientific research. A microscopic impact could put a communications satellite out of action if it hit in just the wrong place.

Similarly, astronauts working in orbit far above the Earth's surface face the danger of micrometeoroids puncturing their space suits. About the worse place an astronaut could be hit is the visor of his helmet, because a crack in the visor would lead to rapid oxygen loss. The space shuttle itself has triple layers in its windows, and the outside component has often needed to be replaced after a flight due to pitting by both meteoroids and man-made space debris.

Apart from having active detectors on satellites, if a piece of hardware has been in space for some time – years, perhaps – it is certain to have been struck many times by minute projectiles, leaving it pitted and cratered. If that hardware can be retrieved and brought back to a laboratory on Earth, scientists can count the perforations it has suffered and look for remnants of the impacting material, which would tell us about its composition; stony meteoroids, nickel-iron grains, tarry stuff from comets, or man-made debris?

Between 1984 and 1990 NASA's Long-Duration Exposure Facility (LDEF) was left in orbit, carrying numerous experimental packages designed to tell us more about the effects of space travel upon the materials from which satellites are constructed. In addition to small solid lumps, satellites are affected by other things such as cosmic rays (charged particles from the Sun and elsewhere in the cosmos), and corrosion by the oxygen atoms that inhabit low-Earth orbit. At the very low pressures of space, plastics tend to lose part of their constituents, and so tend to dry out and become brittle. All these things were studied using LDEF, although in the present context the severe abrasion it suffered due to interplanetary dust during its six years in orbit is most important.

Other space exposure experiments have been flown, not only by NASA but by other nations. Occasionally objects that have spent years in space become available for study even though this was not their original purpose. A particularly good example is the Hubble Space Telescope. During one servicing mission by the space shuttle the solar cells were replaced, and the old ones folded up and brought back to the ground. Their large surface area meant that these were certain to have been extensively pitted by dust grains in space, and scientists were soon poring over them to see what they can tell us about the flux of small particles whizzing past the Earth.

Above An interplanetary dust particle, collected in the Antarctic by melting ice and filtering out the solid grains.
Background An electron microscope photograph of a tiny crater in a space shuttle window. The crater was made by a fleck of paint left in orbit by some previous mission and the window needed to be replaced.
Below Microcraters in the solar cells of the Hubble Space Telescope, returned from orbit after three years' exposure in space. The vertical lines are about 0.04 inches (1 millimeter) apart.

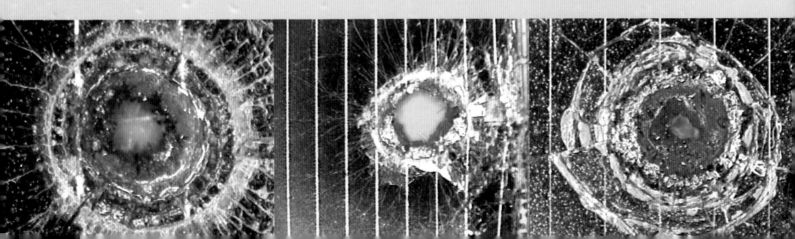

Rocks from the Sky

It was years before modern science would accept that meteorites are rocks from the sky.

Of all the presidents of the United States, Thomas Jefferson could be claimed the most competent on scientific matters. Nevertheless, when two academics from Yale University wrote a report detailing the fall of a small meteorite in Connecticut in 1807, Jefferson is said to have stated that he would "find it easier to believe that two Yankee professors would lie, than that stones should fall from the sky."

We now have many, many meteorites under study, and millions of others await discovery in locations around the world, with no doubt as to their origin. Some come from the asteroid belt, some from the Moon, some from Mars. More questionable is whether any might be fragments of comets. But one thing is sure: we know they fell from the sky, and Jefferson was wrong.

Remember his words, though. Astronomers, myself among them, have been telling the public for years that larger rocks from the sky pose a significant hazard to us all. We are neither joking nor lying.

A Meteorite Smorgasbord

Meteorites come in many shapes and sizes, and not all of them are stones. The first iron to be shaped into knives and jewelry by the ancient Egyptians was meteoritic in origin, and many meteorite samples are composed mostly of that metal. The other major constituent of meteorites is nickel. The reason these metals are dominant is that they are strong, and therefore more able to survive atmospheric entry than weaker elements.

When a saw cuts through an iron meteorite and the flat surface is etched with acid, researchers observe a characteristic crystalline pattern that tells them how the original body cooled in space. In this and several other ways, meteorites provide us with important clues to the formation of the solar system. Apart from anything else, they are the oldest materials on Earth. Radio-isotope dating tells us that they formed about 4.5 billion years ago. All terrestrial rocks are younger than that, because they were melted in the Earth's formation process.

Stony meteorites may be divided into two major classes: chondrites and achondrites. This distinction is based on whether or not the meteorites contain chondrules, roughly spherical structures that range from the size of a pinhead to that of a pea. Chondrules appear to be subunits that formed in the cloud of gas and

Above Excavating a meteorite in the Sahara Desert.
Other photos The variety of shapes, colors, and textures displayed by meteorite samples.

Right The etched interior of a nickel-iron meteorite reveals a pattern of crystals that can tell scientists how the sample formed and cooled from a liquid state.
Opposite Fragments of a stony meteorite found in the Sahara Desert.

dust from which the Sun and the planets gradually accumulated. The chondrules must have grouped together very early on to produce the chondritic meteorites, which are of great scientific interest because they are old and contain some of the earliest material to solidify. It seems that different chondrules formed in contrasting parts of the solar system, so that they provide diagnostic samples of the environment in a range of locations. The achondrite meteorites seem to be more homogeneous in composition, and so they are scientifically less valuable.

Primordial Stuff

Of all the various types of meteorite, the most primitive are the carbonaceous chondrites. These are distinguished from the normal stony chondrites by the large amount of carbon they contain, hence the name. Their carbon is bound up in various organic chemicals; hundreds of different molecules have been identified in these meteorites, including amino acids, some of the basic building blocks of life. Even a passing acquaintance with a carbonaceous chondrite would tell you it is full of organic material: they smell like sulfurous oil.

Such meteorites are weak and brittle, easily breaking apart upon entry. Even if

they reach the ground intact, they are soon lost as rain and other weathering effects rapidly dissolve most of their contents. Either a carbonaceous chondrite is seen on entry and promptly picked up, or it is gone for ever.

This makes these objects very rare, and extremely valuable. If a meteorite is going to land in your garden, then hope it is one of these! Our collections around the world contain fragments of only a handful of carbonaceous chondrites, such as the Allende (Mexico) and Murchison (Australia) falls of the 1960s. There is great excitement in meteorite circles at present thanks to a recent fall in Canada in January 2000.

Above A meteorite encased in ice. After exploding above western Canada in January 2000, hundreds of fragments were sawn out of the frozen surface of the Tagish Lake before the spring thaw sent the rest to the bottom of the lake.

Digging Holes

● **Rocks from space that survive their arrival on Earth are rare.**

The meteorites we discussed on the previous pages must have been saved from oblivion by some chance combination of strength, arriving at speeds close to the minimum possible (around 7 miles or 11 kilometers a second), and a fairly shallow angle of entry. At such an angle the deceleration effect of the atmosphere is more gradual and the energy of the projectile is dissipated before it is entirely burned up, just like a reentering space capsule.

As a result, meteorites tend to land on the ground at very low speeds, much like the terminal velocities of objects dropped from airplanes: perhaps a couple of hundred miles or kilometers an hour, depending on the shape and air drag. Much larger bodies — whole asteroids or comets — punch through the atmosphere and strike the ground at hypervelocity. They are almost entirely evaporated on impact and produce huge craters. Meteorites, though, are traveling much more slowly when they reach the surface, and either remain there or partly bury themselves. The holes they excavate are not much bigger than the meteorites themselves, not ten or twenty times bigger as is the case for explosive impacts on the ground.

Eastern Siberia, 1947

The best-known example of a meteorite fall that produced buried fragments dates back to February 12, 1947. On that day, residents of eastern Siberia were astonished to witness a dazzling bolide streak across the sky, leaving a vast plume of smoke behind it. From Vladivostok it was seen to disappear behind the Sikhote-Alin

hills to the north, and expeditions were soon mounted to find out exactly what had landed. At that time of year the ground was covered in snow, and it was easy to find the debris that had fallen to Earth. Over about a square mile the surface was peppered with holes. Most of them were only the size of a football; others ranged up to 100 feet (30 meters) across.

The largest fragment found of the nickel-iron meteorite responsible weighed in at almost two tons, but thousands of smaller pieces lay around. Many were buried in the snow and not discovered until after the spring thaw. Clearly the incoming object must have weighed many tons, breaking up shortly before reaching the ground because of the shock force imposed upon it by the dense lower atmosphere.

Far left In 1957, the Soviet Union issued a postage stamp to commemorate the Sikhote-Alin meteorite fall a decade earlier, with a painting depicting what onlookers in eastern Siberia saw.

Left The Lonar impact crater in India is one of the few such structures yet identified in Asia. We expect that there are many more awaiting recognition.

Explosions in the Amazon

The Sikhote-Alin bolide is an example of a meteoroid borderline between a crater-forming hypervelocity arrival and a mere meteorite provider. If it had been a bit larger and a bit stronger, it would have reached the ground while still traveling at about 5 miles (8 kilometers) per second. Then it would have exploded, leaving a crater but very little debris. Its main mass would have atomized and been spread by the wind over a wide area.

Similar events may have happened not once but twice in South America in the 1930s. A Catholic missionary sent a report from the upper Amazon region of Brazil to the Vatican that described three separate percussions as a bolide broke up and exploded over the jungle, flattening the forest over a wide area and terrifying the local people. Similarly, an expedition to Guyana organized by the American Museum of Natural History encountered a vast area in which the trees appeared to have been snapped off halfway up the trunk. Locals' descriptions suggest a phenomenal fireball was responsible.

Wabar Craters, Saudi Arabia

Deep in the legendary Empty Quarter of Saudi Arabia – the Rub' al-Khali – lies an area more than 100 acres (37 hectares) in extent, covered with black glass, white rock, and iron shards. It was first described in 1932 by Harry St. John "Abdullah" Philby, a British explorer perhaps better known as the father of the infamous Soviet double agent, Kim Philby. The site he depicted had been known to several generations of nomadic al-Murra Bedouin as al-Hadida, "the iron things."

Expeditions in recent years have shown that apart from two large craters, at least one smaller bowl also exists, but all are rapidly

Below A beautiful sample of natural glass found in the western desert of Egypt, produced by the melting of sand in a cosmic impact.

Shatter Cones

Hypervelocity impacts like that which excavated the Lonar crater in India, pictured opposite below, do more than simply dig a large hole in the ground. The extreme transient pressures experienced by the target rock cause various changes in the rock's structure that allow the occurrence of an impact to be deduced long after the crater itself has eroded away. Minerals such as coesite and stishovite are formed that are diagnostic of an impact event.

Shatter cones, like those pictured here (note the rock hammer for scale), are another characteristic feature of rocks that have suffered an impact. The target rock is cracked and faulted in a pattern that indicates not only that a phenomenal shock wave passed through the rock, but also the direction of the wave's propagation.

being filled with sand. This allows us to date them to a maximum age of a few hundred years, and it seems likely that they were formed following the sighting of a bright bolide over Riyadh just over a century ago. The metallic impactor, whose fragments had been slowed down by atmospheric drag to about half their initial speed by the time they hit the dunes, is estimated to have measured about 10 meters or yards across. Exploding on contact, the energy released was equivalent to about 100 kilotons of TNT. The extreme radiant heat generated by the impact melted much of the surrounding sand, forming a type of natural glass called impactite. This material, a classic "fingerprint" of a high energy impact, is found scattered around the site. Impactite also occurs in places where no craters are known, such as Egypt's western desert. Presumably, shifting sands have covered the craters – as is happening at Wabar – leaving only traces of glass to hint at past calamities.

Larger Lumps

Our atmosphere is deceptive. It shields out most small projectiles, making us think we're safer than we really are.

It may come as a surprise to know that the thin gaseous shroud around the Earth stops most asteroids smaller than 200 feet (60 meters) from reaching the ground intact. Because of their phenomenal arrival speeds, these rocks disintegrate and burn up at altitudes of 5 miles (8 kilometers) or more. For cometary material – made of a mixture of ice and smaller stony components – theoretical modeling indicates that even a lump 500 feet (150 meters) in dimension will not penetrate deeper than a height of around 10 miles (16 kilometers). This does not mean that these projectiles are not dangerous, however. When they explode they produce blasts whose shock waves propagate rapidly downwards, flattening everything for miles

around. But it does imply that impact craters will not be formed.

Tunguska River, Siberia, 1908

What happens when a rock about 200 feet (60 meters) across enters the atmosphere at 10 or 20 miles per second? In 1908, the inhabitants of central Siberia found out, to their cost.

At about 7:17A.M. on June 30, the nomadic reindeer herders who populate the mosquito-infested swamps and forests of the Tunguska River region – terrain known as the *taiga* – were already going about their business when the sky was suddenly lit up by a massive fireball approaching from the southeast. It grew in intensity until its radiance set the forest ablaze, and even burned the shirts off some people's backs. When it reached an altitude of about 5 miles (8 kilometers) it abruptly exploded into billions of smaller fragments that burned away to nothing. The energy of the blast has been calculated at the equivalent of between 10 and

Above A postage stamp issued by the Soviet Union in 1958 to mark the fiftieth anniversary of the Tunguska event. On the right is Leonard Kulik, who led the first expeditions to the site from 1927 onward.

Below and left Photographs of the fallen trees at Tunguska taken in the 1930s, when the first scientific expeditions came to investigate what had happened a couple of decades earlier.

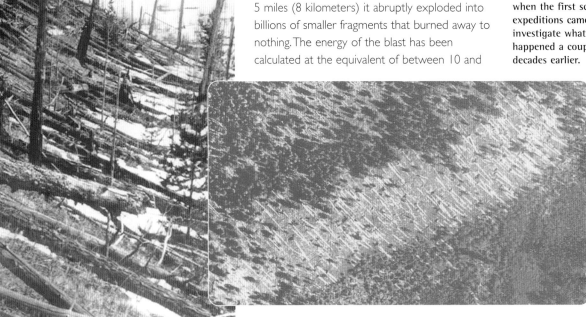

Left The charred and fallen trees at Tunguska are still obvious today, as seen in this photograph taken in the late 1990s.

20 megatons of TNT – a thousand times the power of the Hiroshima bomb. The trees burst into flames over an area of about 1,000 square miles around the epicenter. Thirty seconds after the explosion, the blast wave – moving at the speed of sound – reached the surface. This flattened the trees as if they were matchsticks, seconds before a phenomenal wind extinguished the flames like a giant puff of breath.

Residents living many miles away reported instant chaos. They told of houses being blown down, and of people and animals being swept away by the wind. Some hundreds of miles distant the drivers of the trans-Siberian railway stopped their locomotive as the ground seemed to shake. The jolt was registered by seismometers in Britain, and the atmospheric shock wave, detectable by sensitive barometers in other continents, circled the Earth twice before it was spent.

Later expeditions would find scenes of unimaginable devastation, and although scientific parties didn't visit until 1927, even two decades on there was ample evidence of what had happened. The trees had been mostly stripped of branches, and charred only on the side facing the bolide's path. Early investigators were able to determine the epicenter of the blast by plotting the directions of the fallen trees, radiating away from ground zero. Here they searched long and in vain for the crater they expected to find. They knew enough to suspect that a cosmic impact had been responsible, but the realization that the Earth's atmosphere fragments and dissipates large solid rocks was still decades away.

There are some small impact craters on Earth – Wabar, the Henbury craters in Australia, Odessa in Texas, and Meteor Crater in Arizona – but these were all formed by iron asteroids. Rocks of the same size don't reach the ground intact, so they leave scant, short-lived evidence of the danger they pose. The Tunguska event is merely a recent example.

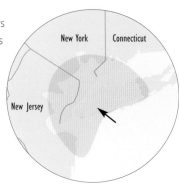

Below The trees were blasted flat at Tunguska over an area of about 800 square miles (2,200 square kilometers), in a butterfly pattern symmetric about the projectile's arrival from the southeast, as shown here. The area in question is overlaid on a map of the New York City metropolitan area to give an idea of the extent of the devastation.

Not a New Idea

Today most of us are familiar with the idea that the dinosaurs were killed off 65 million years ago following the impact of a massive asteroid or comet. But how many of us are aware that this theory is anything but new?

In truth this notion – that the terrestrial environment is upset occasionally by unwelcome cosmic visitors – has been around for centuries. The record shows it resurfacing periodically in the writings of some great scientist before it is once again hushed by the establishment.

Seventeenth-Century Comets

When Edmond Halley was born, the origin and nature of comets were totally unknown. Some thought they were temporary globules expelled by the planets; others believed they were visitors from interstellar space.

Late in 1680, two bright comets were seen some weeks apart. They were assumed to be two distinct bodies, but the first Astronomer Royal, John Flamsteed, investigated their paths and showed that they might have been two views of the same comet, before and after its closest approach to the Sun. In effect, Flamsteed was suggesting that comets are regular members of the solar system and that they orbit the Sun as the planets do.

Flamsteed's suggestion was vehemently disputed by Isaac Newton, which is perhaps surprising in light of the work Newton was then doing on his own theory of gravitation. But eventually Flamsteed was proved right, and within a few years Halley showed that another comet seen in 1682 could be identified with previous apparitions in 1456, 1531, and 1607; this is the comet that bears his name. Newton recanted and brought elliptical cometary orbits into his gravitational theory which was published in 1687.

This made it obvious that comets could hit the Earth, an uncomfortable notion at the time for both the political and religious establishments. Halley got into further trouble for suggesting that impacts might have caused catastrophes such as the biblical flood. Rather irrationally, Newton described cometary interactions with our planet as benign, and that they simply brought new life and beneficial substances with them.

The Idea Resurfaces

Newton moved to London and William Whiston took his position at Cambridge University. Whiston was less concerned to avoid upsetting the clerical authorities of the day. Impressed by Halley's ideas, in 1696 he published *A New History of the Earth*, in which he tried to explain on a scientific basis various phenomena described in the Bible. The work was a great popular success, but drew much criticism from the Church.

We now know that the vast coma of a comet is actually a very tenuous vapor, but at that time it was thought to be largely solid, implying a huge mass. Based on this mistaken belief, Whiston suggested that a very close passage by a comet had cracked the Earth's crust, releasing subterranean waters and also attracting a tide some miles high. This, he said, was how Noah's flood originated. He even suggested a date for the event, estimating it at 2349 B.C.

Whiston went on to suggest that a comet would cause the end of the world in A.D. 2255. Halley had incorrectly derived an orbital period of 575 years for the comet of 1680, and Whiston claimed that on its next return it would pass close by again. Either the comet would hit the Earth, or it would drag us from our orbit, plunging our planet dangerously close to the Sun where it would fry in the solar blaze.

Whiston's analysis may have been wrong, but his crime in the eyes of the authorities was to have made these dangerous ideas public. He soon lost his job.

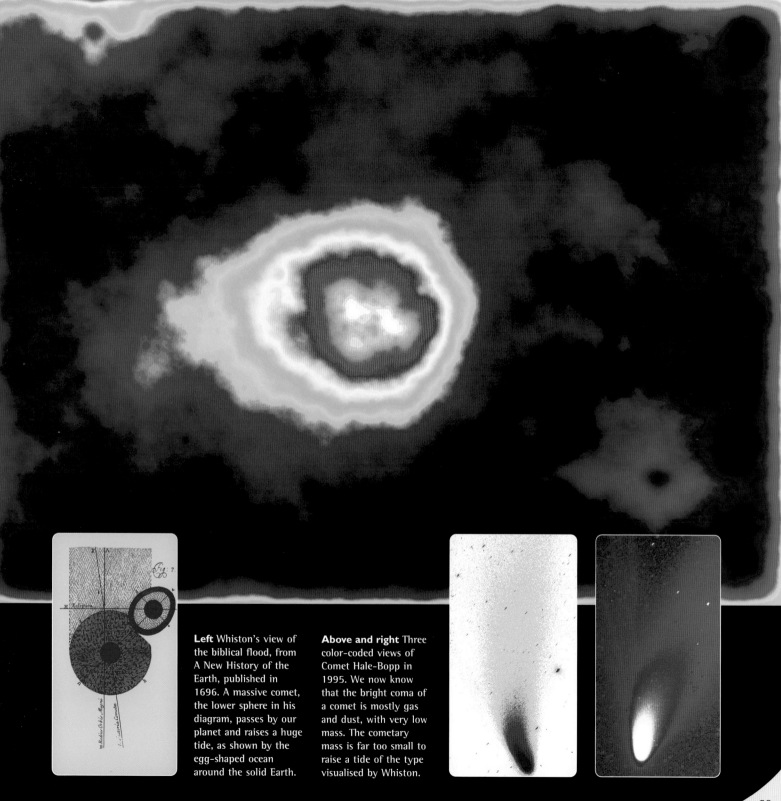

Left Whiston's view of the biblical flood, from A New History of the Earth, published in 1696. A massive comet, the lower sphere in his diagram, passes by our planet and raises a huge tide, as shown by the egg-shaped ocean around the solid Earth.

Above and right Three color-coded views of Comet Hale-Bopp in 1995. We now know that the bright coma of a comet is mostly gas and dust, with very low mass. The cometary mass is far too small to raise a tide of the type visualised by Whiston.

Catastrophism

Did the gross environmental changes evidenced in the geological and paleontological record occur gradually, or were they the result of sudden calamities?

In the eighteenth century, the majority of astronomers still held to the belief that comets were actually vast masses even heavier than the planets. The sheer physical dimensions of the bodies concerned were responsible for this erroneous view – the coma of a comet, which makes it very bright in the night sky, may be more than ten times the terrestrial diameter, and even bigger than Jupiter. But the truth is that the coma has a very low density, and almost all the mass of a comet is contained in the solid nucleus, with a typical size of around a mile.

The notion that comets might occasionally collide with other celestial objects, such as the planets or the Sun, had already gained a foothold. In 1745 the French nobleman Count Buffon – a pioneer of planetary science – suggested that the planets had actually been formed from the rebounding mass of material

Left A nineteenth-century French cartoon showing the Earth being torn apart by a comet while a laughing Moon looks on. **Background** The different forms that comets may take, according to the first century Roman writer Pliny.

Below A Turkish view of the great comet of 1577.

ejected after a huge comet had plummeted headlong into the Sun.

Buffon's hypothesis was wrong, but it does show how people were thinking at the time, especially in France. In the late eighteenth century the Marquis de Laplace, whose name appears in many fields of mathematics and physics, revived Whiston's idea of a giant comet passing close by the Earth. Laplace wrote:

The seas would abandon their ancient positions, most of the human race would be drowned in the universal deluge, entire species would be annihilated if a comet approached us.

It was in this climate that Baron Cuvier, one of the founders of paleontology, excavated parts of Paris as it was reconstructed after the French Revolution. He thought of the biblical flood as having been simply the most recent in a series of catastrophes that had wreaked havoc upon the Earth, and he found evidence for earlier extinctions in the strata beneath the city.

He would identify fossils of sharks and other marine species in one layer, and deer and other

land-going animals in the next layer. In all, he delineated six calamitous events, with each one followed by the proliferation of new species.

Lord Byron

Cuvier represented what is known as the Catastrophist School, which saw the Earth's history as a series of little-changing eras punctuated by massive upheavals. Although this school of thought was not limited to mainland Europe, British science had to some extent rested on its laurels after Newton. It was almost as if there was a belief that all that needed to be done had already been done. Outside of scientific circles, however, some mavericks realized that new philosophies were emerging elsewhere, and they traveled extensively to discuss these matters in the salons of the European capitals.

One of these mavericks was Lord Byron, the poet. He maintained that the human race was merely the present occupant of the top rung in the hierarchy of beasts, and believed that previous incumbents had been obliterated by comets that had slammed into the planet. This, remember, was before we knew of behemoths like the dinosaurs. In the early 1820s, Byron stayed a couple of years in Pisa, and spoke of his theories concerning the fossil record.

Astonishingly, Byron realized not only that comets had struck the Earth in times past, and that they would do so again, but he also foresaw the possibility that the human race could do something about it. Since this was the age of the Industrial Revolution, with steam powering the wonderful new machinery that made manufacturing so efficient, Byron thought only in terms of the capabilities of steam. But he did suggest that we could stop a comet on a collision course, and avert our doom. To that extent, Byron may be counted as one of the originators of the Spaceguard concept.

Above and background A depiction of the spectacle of Donati's comet over Paris in 1858.

The Two Charlies

While Byron was absorbing foreign notions of comets as agents of catastrophe, the opposing Uniformitarian School was expanding its hold in Britain. This viewpoint saw change as gradual, brought about over very long periods by the natural forces that may be seen in action every day. There was no need, in this philosophy, to appeal to calamities such as comet impacts causing sudden and widespread change.

It reached its apex through the work of Charles Lyell, who published his *Principles of Geology* in 1830. He argued that all the features seen in the geology of Europe could be explained by the slow action of weak changes over extensive time spans. He explained Cuvier's distinct fossil beds as regional discontinuities caused by factors like local variations in climate, rather than global disruptions.

Lyell's influence spilled over into other areas of science. In formulating his hypothesis of evolutionary change in zoology, Charles Darwin applied uniformitarian thinking. He argued that the forms of related species change over extended periods through the gradual action of environmental pressures. Even today many evolutionists see the basic tenets of Darwin's *Origin of Species*, published in 1859, as being correct in detail; whereas it seems that evolution proceeds in fits and starts rather than as a slow and gradual process. Darwin was not thinking about massive impacts wiping out many species, leaving niches free for the rapid development of others that are favored purely by chance. Remarkably though, Byron had correctly guessed – several decades earlier – that this is exactly what must happen.

Advances in America

As the nineteenth century progressed, the idea that cometary impacts were both real and dangerous slipped out of the mainstream.

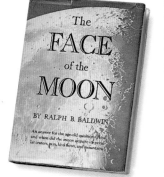

Above As early as the 1940s books were published claiming that asteroids and comets pose a threat to life on Earth.

In remote spots a few free thinkers had other ideas. In New Zealand in the 1880s a professor of physics, Alexander Bickerton – teacher of Lord Rutherford, the man who split the atom – wrote about massive impacts shaping the face of the planet. But he was ignored.

The United States was one of the few places where the effects of impacts were evaluated. Previously we heard about the work of Daniel Barringer at Meteor Crater, and how mainstream geologists opposed him in his interpretation of that structure as the result of an asteroid impact. Even in the 1920s the notion that huge rocks from space might smash into the Earth and leave holes in the ground was considered fanciful. The 1930s were to be a watershed.

Harvey Nininger

Harvey Nininger was the father of American meteoritics, amassing a fabulous collection of these space rocks. But he realized that larger objects must land at high speed, destroying themselves in the process while excavating vast craters. He visited the 180-yard (165-meter) wide Odessa crater in Texas, noting its shape and the meteoritic iron shards all around, and he left in little doubt about its origin. Around the same time, in the early 1930s, the somewhat similar Henbury craters in Australia were described in scientific papers, and Philby gave an account of the Wabar craters in Saudi Arabia.

Nininger led expeditions to other American craters, such as the tiny Haviland depression in Kansas. His work ensured that the reality of impact cratering would be taken seriously. Perhaps his most important contribution was the realization that the asteroids found on Earth-crossing orbits in the 1930s would cause global devastation if they were to run into our planet, as some inevitably must do. In 1941 he published an account of what would happen when a decent-size asteroid hit, and suggested such events were the cause of the discontinuities in the geological and paleontological record. These are the boundary events that Cuvier recognized in the rock strata below Paris.

In essence, Nininger saw asteroid impacts as the origin of the mass extinctions that have punctuated the development of life. He did everything except mention the death of the dinosaurs directly, but his all-encompassing theory viewed large impacts as a major driving force behind the evolution of species.

Watson and Baldwin

Despite Nininger's work, there was still a broad lack of understanding about the origin of the lunar craters. Two popular books soon brought

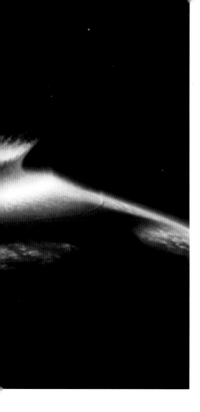

Left American artist Don Davis's impression of a massive asteroid hitting the Earth. Although such impacts did occur early in our planet's history, there are no such vast asteroids left now. But much smaller objects can still cause mass extinction events.

the latest information into the public domain. The first of these, *Between the Planets* (1941), by Fletcher Watson, masterfully described the debris orbiting the Sun (asteroids, comets, and meteoroids), and told how several small terrestrial meteorite craters had recently been recognized for what they were.

The second was *The Face of the Moon* (1949) by Ralph Baldwin. The real story behind Baldwin's book was the recognition that much larger craters must exist on the Earth than had previously been found, and that they would be linked to widespread cataclysms. Baldwin hinted that the vast craters on the Moon must be echoed on our own planet:

The explosion that caused the crater Tycho would, anywhere on Earth, be a horrifying thing, almost inconceivable in its monstrosity.

Baldwin did not stop short of sharing the implications of recent discoveries with his readership:

Background A fanciful depiction of a huge spherical asteroid heading for our planet, from Kelly and Dachille's *Target: Earth.*

Tiny asteroids, each of which could, in some future year, entirely devastate an American state or a European country; tiny asteroids which might wipe out local species of flora and fauna. Sudden disappearances of long-established groups of contemporary life have been recorded in past geologic history. Is it not possible that the causes of these occurrences were meteoritic impacts?

Kelly and Dachille

Baldwin's book brought the idea of an impact hazard to a wide audience. Although his warnings were ignored in most spheres, they did attract some attention. In California in 1953, Allan O. Kelly and Frank Dachille published *Target: Earth.* They wrote about the dinosaurs being wiped out by an asteroid impact, although their physics was a little bit awry (they saw the Earth's spin axis being shifted, whereas this is not a feasible consequence of an impact). What Kelly and Dachille said about the need for a defense system against impacts is interesting, however, especially because they were writing before the dawn of the Space Age:

Therefore, it behoves us to consider ways and means to ward off a Day of Reckoning that may be set up in the mechanics of the Solar System and the Universe. In the increasingly numerous scientific discussions of rockets for interplanetary travel, and in the consideration of man-made satellites or artificial moons, we see the beginnings of a system for the protection of the Earth. This system will require perpetual surveillance of a critical envelope of space with the charting of all objects that come close to a collision course with the Earth. It will require, further, that on the discovery of a dangerous object, moves be made to protect the Earth. To this end might be used rocket "tug boats" sent out to deflect and guide the object from the collision course.

This was farsighted indeed.

The Death of the Dinosaurs

Between the 1950s and 1970s, a handful of researchers argued that impacts must affect life on Earth.

Astronomers looking outward had spotted asteroids and comets, and realized that some must slam into the Earth from time to time, with cataclysmic consequences. Geologists and meteoriticists had found crater evidence for such events on Earth and on the Moon. What were the implications for life?

In the 1950s, the handful of physical scientists working on the subject had dared to suggest that the flora and fauna of our own planet might have been severely affected by massive impacts, some going as far as to say that asteroids and comets were responsible for the mass extinctions that pepper the paleontological record. How did the paleontological profession react to this trespassing on their territory?

The answer is either not at all, or largely in a negative way. Even today only a minority of paleontologists incorporate impacts into their worldview, and in the past the response was even more hostile. Nevertheless, in the mid-1950s, M.W. De Laubenfels of Oregon State University suggested somewhat tentatively in the *Journal of Paleontology* that asteroid impacts might have killed off the dinosaurs. When he learned about the Tunguska event of 1908, he scaled up to larger impactors but still did not imagine the full ferocity of a major impact.

Impacts: How Often?

It's all very well to nominate impacts as a primary cause of mass extinctions because such a scenario is possible, but is it probable? The Earth is a relatively small target. How often can major impacts occur?

You can reckon a crude probability of an impact by simply noting that the cross-sectional area of our planet is about one part in two billion of the area of a sphere with radius equal to our distance from the Sun. An Earth-crossing asteroid or comet must pass through that sphere twice in each orbit, once on the way in, once on the way out. That gives the probability of a hit equal to one in a billion per orbit. That's a lower limit, however. Most asteroids have orbits that stay close to the plane of Earth's orbit, and so they must cross that imaginary sphere rather close to our path, making an impact more likely.

The person who revolutionized our understanding of the frequency of impacts was Ernst Öpik, an Estonian who worked for many years at Armagh Observatory in Northern Ireland, and at the University of Maryland. In the 1950s he not only derived ways of calculating

Above Ernst Öpik, who first worked out how often asteroids and comets must strike the planets, worked at Armagh Observatory in Northern Ireland. This continues to be an important center for work on the subject. **Opposite left** Moon rocks like this one returned in the Apollo program show clear evidence of the shocks produced in impacts.

realistic impact rates, he also indicated how the phenomenal explosions involved could cause extinction events. It was Öpik who realized that Mars and Mercury must be pockmarked with craters – a fact that was established before he died in 1985.

Confirmation from Apollo

Despite these undeniable strides forward, many still dismissed impacts as the origin for the lunar craters throughout the 1960s. The problem was part psychological: if you accept that asteroids and comets crater the Moon, then it follows that the Earth must also get hit, which is an uncomfortable thought.

The final acceptance of impacts as the cause of lunar craters did not come until the first moon rocks were brought back in the Apollo project, showing unarguable evidence of their history. The Moon is struck frequently by cosmic projectiles, which excavate huge craters many miles across. The Earth must be too.

Tektite Evidence

The next step forward in linking cataclysmic impacts with faunal extinctions came in 1973. Harold Urey, a pioneer of planetary science and Nobel Prize-winning geochemist, noted in the influential journal *Nature* that the ages of the major tektite concentrations seemed to be coincidental with the dating of mass extinctions. Since tektites are formed by massive impacts, the implication is obvious: the impacts caused the extinctions. This should have been a pivotal juncture, but Urey's paper was all but ignored by most researchers. The widespread acceptance of impacts as the origin of extinctions would have to wait for some years.

Above The Kara-kul crater in Tajikistan, formed 25 million years ago, is 32 miles (52 kilometers) across.

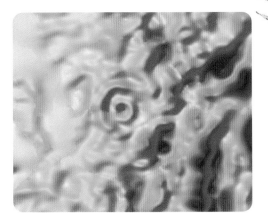

Above and left The recently discovered Woodleigh crater in Western Australia, which is at least 80 miles (130 kilometers) across, may be linked to the Permian-Triassic mass extinction event 250 million years ago.

Near Misses and Bull's-Eyes

In the late 1970s, British astronomers linked comets to upsets of the terrestrial environment.

Probability calculations indicate that a large comet might hit the Earth every ten million years or so, whereas asteroids (because of their larger number) should produce more frequent collisions. It all depends on the threshold size one takes to be of interest.

As far as we can estimate, an asteroid a half mile (800 meters) in size arrives about once every one hundred thousand years. This may be large enough to cause a global catastrophe – a perturbation of the environment sufficient to cause the death of a large fraction of mankind – but the actual threshold may be closer to a mile.

The threshold size for the impactor may be estimated by scaling up from the area of damage caused by nuclear bombs: given that a one megaton bomb devastates a certain number of square miles, how many megatons would lay waste to a whole continent, upsetting the global environment in the process? The answer is about a million megatons, which corresponds to the impact of a one-mile asteroid.

The most significant killing mechanism would

Right Comet Hale-Bopp in 1996. Rather than be hit by the solid nucleus itself, it is much more likely that the Earth will pass through the dusty coma of a comet.

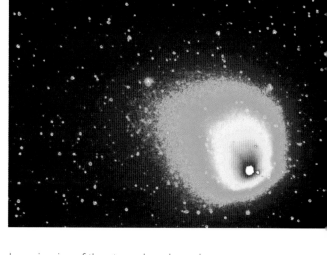

Below Scottish astronomer Bill Napier recently wrote a best-selling novel on the danger of impacts.

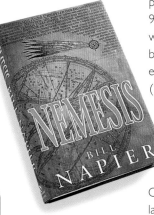

be poisoning of the atmosphere by various noxious chemicals produced in the explosion. This, though, would not be sufficient to cause a mass extinction event. Extinction means a 100 percent death rate for a species, no less. While 99.9 percent kills most individuals, the species would soon recover and the hiccup would not be noticed in the fossil record. To get a mass extinction event an impactor perhaps 10 miles (16 kilometers) in size would be needed, releasing energy equivalent to a billion megatons of TNT.

But there are no known Earth-crossing asteroids this big. There are some that pass close by, such as asteroid Eros, and their orbits can alter to make a collision feasible. Comets Halley, Swift-Tuttle, or Hale-Bopp are large enough to release that order of energy, but such big comets are rare. Maybe an actual impact is not necessary. Would a near-miss suffice?

Cometary Clouds

Although an impact by a cometary nucleus is very improbable, the coma or cloud of gas and dust around that nucleus is huge, perhaps 100,000 miles or more across. This is why comets appear so bright in the sky, because the

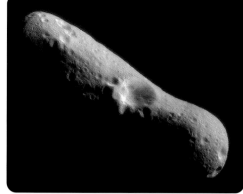

Left Very large asteroids like Eros would need to hit the Earth to cause a mass extinction event. But could comets cause upsets in other ways?

cloud reflects so much sunlight. To put it into perspective, the coma may be as big as the moon's distance from the Earth. Which means that a comet passing closer to us than the lunar orbit would envelope our planet as it moved by, and such events must have occurred fairly often, probably several times in each million years.

Sir Fred Hoyle is one of the doyens of British astronomy, having made many seminal contributions to astrophysical theory. With his colleague Chandra Wickramasinghe he has also argued that comets are fundamental to life and death on Earth, seeing them as bringing organic chemicals and maybe life itself to our planet. In 1978 Hoyle and Wickramasinghe proposed that the dinosaurs were killed not by a collision but by the close passage of a comet near Earth, the massive coma depositing so much dust in the upper atmosphere that the climate cooled substantially. An Ice Age ensued, which the dinosaurs were ill-equipped to withstand. Apart from being unable to control their own body temperatures, the reptiles would have starved to death as the food chain collapsed.

The Big Picture

In recent years Hoyle and Wickramasinghe have collaborated with two other British astronomers, Bill Napier and Victor Clube, on investigations of how comets and the dust they produce might be major perturbers of the terrestrial climate. In the late 1970s, though, Napier and Clube were working on their own model for the death of the dinosaurs, publishing in 1979 in the journal *Nature* a discussion of a general theory for terrestrial catastrophism.

Napier and Clube noted Urey's link between tektites and mass extinctions, implicating gargantuan impacts by asteroids and comets. They then considered the ages of known impact craters and found that these again provided evidence of a link. The big advance made

Above, clockwise from bottom left Sir Fred Hoyle, Chandra Wickramasinghe, Bill Napier, and Victor Clube have suggested various mechanisms through which comets can affect life on Earth, and also linked the movement of the solar system through the galaxy to epochs of enhanced numbers of comets.

possible by Napier and Clube was that they saw the impacts as being the result of enhancements in the numbers of comets entering the planetary region. They linked this increase in the comet population to the movement of the solar system around the galaxy. At that stage the hypothesis was based on the way in which the Sun moves inward and outward from the galactic center. Later models have looked instead at the way our solar system oscillates up and down through the galactic plane, repeatedly pulled back by the mass of stars in that plane.

The idea that events on Earth could be linked to the dynamics of the galaxy as a whole was a radical concept. Its basis had been mooted before, with scientists wondering why the mass extinctions and geological boundary events appear to be cyclic. But the gathering together of the physical evidence put this question firmly on the agenda for the first time.

Hard Evidence

Earlier suggestions of a link between massive impacts and mass extinctions lacked solid evidence. But proof finally appeared two decades ago.

Above The team that uncovered the vital clue to the dinosaurs' demise. From left to right: Helen Michel, Frank Asaro, Walter Alvarez, and Luis Alvarez.

As far as most people are concerned, the beginning of the idea that the dinosaurs were killed by an asteroid became public knowledge in 1980. Then, accompanied by much media hype, an article was published in the journal *Science* that revealed hard evidence that a massive asteroid hit the Earth 65 million years ago. That is, at exactly the time that the dinosaurs and many other animal families disappeared in an apparently brief event.

The Iridium Layer

The brevity of the transition from the Cretaceous era to the Tertiary era – the so-called K/T boundary event where the dinosaur fossils suddenly cease – is what had initially stimulated the interest of a team from the University of California at Berkeley. Walter Alvarez, a geologist at the university, wanted to know how long the transition took to complete. His father, Luis, was a Nobel Prize-winning physicist. Together they thought they might use the rare element iridium to delineate the time scale of the boundary, where one type of rock abruptly stops and another begins.

The K/T boundary occurs worldwide, so it is not the result of some simple local change, such as a river washing away the upper strata and

then laying down new rock above. There was some radical alteration in the global environment at that time. But how long did this upset take?

The Alvarez team reasoned that although there is little iridium on Earth, it is plentiful in meteorites and interplanetary dust. Based on the assumption that the influx of this dust is constant, the concentration of iridium in the strata would be an indicator of how long the sedimentation process that produced the rocks had continued. If the sedimentation were slow, there would be a greater amount of iridium.

When the iridium level was traced, the result was surprising. There was a huge spike in the amount of this element present in the rock right on the K/T boundary. The rare metal osmium showed a similar enhancement. They had to conclude that their original assumption was incorrect. Instead of representing a steady and

Below Exposed rock strata in Texas show a radical and abrupt change in color from the Cretaceous era (lower down, and so older) to the Tertiary (not so deep, and so younger). All dinosaur fossils, and many other species, suddenly halt at the junction between these eras.

NOT A NEW IDEA

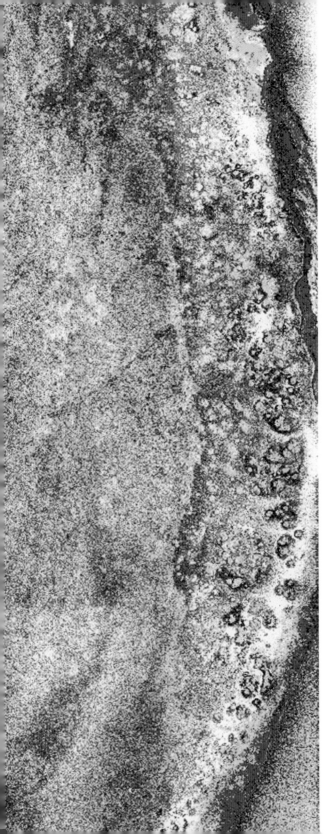

Left In 1990, a huge crater on the Yucatán peninsula in Mexico was identified as the result of the impact 65 million years ago. This recent image of a section of its rim, very difficult to recognize on the ground, was constructed using radar data collected from the space shuttle.

gradual influx of dust, the iridium in the layer must have resulted from a single massive injection of extraterrestrial material.

The Alvarez team's working hypothesis, based on the amount of iridium and its known concentration in meteorites, was that an asteroid between 5 and 10 miles across had struck the Earth and mostly vaporized. Its constituent materials were then distributed all over the globe in a thin layer, which could be detected anywhere the K/T boundary stratum was accessible. All of a sudden the idea that such an impact had caused the biotic extinction at that point in our planet's history was inescapable.

Above It has happened before. When will it happen again?

Chapter 6
Are We Sitting Ducks?

The dinosaurs did not see their nemesis coming. And even if they had been dimly aware of a comet in the sky, brightening from one night to the next as it neared the Earth, there is nothing they could have done about it. But we have intelligence, science and technology, and opposable thumbs.

We are not sitting ducks. We could spot an asteroid or comet destined to hit us if we bothered to look for it. It is also possible that we could divert such an object, and thus avoid the fate of the dinosaurs. But the first step is the looking.

Found by Accident

The development of new astronomical instruments has progressed by leaps and bounds since Galileo turned his first rudimentary telescope toward the sky in the early seventeenth century. Advancing technology allowed the building of ever larger telescopes throughout the 1800s. The first half of the twentieth century saw the construction and the first 100-inch aperture systems. Before long 200-inch (5-meter) telescopes were brought into service.

Because conventional astronomical telescopes are designed for looking at just one star or galaxy, they cover a very restricted area of the sky. A telescope that can cover an area equal to that of the Moon would require 100,000 shots to photograph the celestial hemisphere. And since most telescopes cover much smaller areas than that, a million or more images would be required to do the job.

The problem with asteroids, though, is that thay can appear anywhere in the sky. Although main belt asteroids are mostly restricted to a band within about 20 degrees of the ecliptic (the apparent path of the Sun and planets across the sky), near-Earth asteroids may approach from any direction. If your photographs are only covering a millionth of the sky then your chances of catching one are slim.

All this changed in the 1930s with the introduction of the first Schmidt telescopes. These cameras cover a much larger area than a conventional telescope, and produce an undistorted image by employing a complex combined lens and mirror optical system. This made it possible to photograph a wide region of the sky at one time, typically 6 degrees across. Although this is still only one part in a thousand of the total sky area, it did enhance the chance of detecting near-Earth asteroids by a factor of a thousand. And soon several were discovered quite by accident, as astronomers were photographing the sky for other reasons.

Project Icarus

The first three Earth-crossing asteroids were detected in the 1930s; a few more followed in the 1940s. One of these was 1566 Icarus, named for the character from Greek mythology. The story tells of how Daedalus and his son Icarus escaped their prison tower in Crete by fashioning wings from feathers and wax. Icarus failed to heed his father's warning and flew too close to the Sun. The wax melted, and he fell into the sea and drowned. Asteroid Icarus was quickly recognized to have an orbit that takes it closer to the Sun than Mercury, hence its name.

From time to time this object comes fairly close by the Earth, and it caught the popular imagination during the 1950s and 1960s each time someone suggested that a collision with the Earth was a possibility. In 1967 it became the focus of a class project at the Massachusetts Institute of Technology where students were given the task of inventing a way to divert it, with only 15 months' warning of an impending impact. Their solution made use of the Saturn V boosters then being built for the Apollo program – six launches carrying a series of 100-megaton hydrogen bombs to intercept the incoming asteroid. The result was Project Icarus, the title of their report published for all to peruse. It's still an interesting and instructive document to read.

Left Asteroids and comets look different through a telescope. At far left is the trail of an asteroid moving between a background of stationary stars. Near left, a comet produces a fuzzy image due to its surrounding cloud of gas and dust.

Above The 1,000-foot (300 meter) radar system at Arecibo in Puerto Rico has been a vital tool in studies of near-Earth asteroids and comets.

Finding the Vermin of the Sky

How can astronomers identify faint, fast-moving asteroids among the panoply of stars?

We read earlier that some astronomers regard asteroids as little better than vermin because of their tendency to spoil photographic images by producing streaks in the field of view of the telescope as they move through space.

Others desperately want to find them. To us, the stars are simply background markers against which we can measure an asteroid's position and determine its orbit. The first near-Earth asteroids were found by accident. But all the time awareness was growing that these objects could present a real threat to life on Earth, and the 1970s saw the start of the first projects dedicated to searching for them.

Gene Shoemaker, Polymath

The initiator of much of the work in this area was Eugene Shoemaker – "Gene" to his many friends. Shoemaker trained as a geologist, and always had a special interest in impact craters. From the late 1950s he made detailed studies of Meteor Crater, and he was instrumental in setting up the Branch of Astrogeology of the U.S. Geological Survey, in nearby Flagstaff, Arizona. He made his home there with his wife and scientific partner, Carolyn.

Shoemaker was at the forefront of the Apollo project throughout the 1960s, and would surely have made a trip to the Moon himself if it weren't for a medical condition. As it was, he trained the astronauts who did visit the lunar surface in what to look for, and told them which rocks to collect. When the Apollo project came to an end in 1972, Shoemaker continued his geological investigations of terrestrial impact craters, but he also turned his eyes skyward, starting a search for the asteroids responsible.

Above Comet Shoemaker-Levy 9 soon after its discovery in 1993.
Top Impact scars on Jupiter photographed with the Hubble Space Telescope after SL9 slammed into the planet in mid-1994.
Background A color-coded image of Comet Hale-Bopp, showing the vast expanse of its dust and vapor cloud, through which background stars can be seen..

Up until then, asteroids had been spotted by the trails they left on long-exposure photographs. During a one-hour exposure an asteroid might move 50,000 miles (80,000 kilometers) across the sky, leaving a trail the size of a match head on the photographic plate. Its orbit would then be calculated from subsequent observations.

Shoemaker decided on a different method. He took brief-exposure shots of the night sky using a Schmidt camera located at Palomar Observatory in California. By taking pairs of photos separated in time by an hour or more, any moving dot could be identified and followed up. Because he used shorter exposures, each lasting perhaps only a few minutes, much more of the sky could be covered in a given time period, and many more asteroids found.

His program was extremely successful and hundreds of asteroids were discovered, including dozens of Earth-crossers and many comets. He was later joined by amateur astronomer David Levy, already an established comet discoverer, who scoured the skies from his home near Tucson, Arizona. Each month the three of them – Gene, Carolyn, and David – would drive to southern California for a week of observations at Palomar, rarely failing to find some remarkable object that would subsequently be followed up and tracked by other interested astronomers.

The Discovery of SL9

They had already discovered many long-period comets, each called Shoemaker-Levy with some official designator to differentiate them, as well as eight short-period comets similarly named with a numeral after them. But in early 1993 they picked up a very unusual object, designated SL9, that was soon headline news.

All their other comets were in Sun-centered orbits, but this time they were looking near Jupiter, and their new discovery turned out to be in orbit around that planet. As if that were not unusual enough, SL9 presented a bizarre profile. The initial photographs they obtained showed it to be bar-shaped, with an elongated nucleus, and quite unlike any comet that had been seen before. Turning to other astronomers for confirming observations, it soon became clear that SL9 was a comet that had broken up into a string of fragments. As we know, these fragments plummeted into Jupiter the following year, as the world looked on in wonder.

Gene Shoemaker was already famous among planetary scientists as the father of modern impact cratering studies. Now he became a household name. For more than a decade he had made expeditions to Australia with Carolyn, mapping the many craters there. In July 1997 their car suffered a head-on crash with another vehicle in the Tanami Desert. Gene was killed instantly. Carolyn survived.

Above Astronomer David Levy with his array of telescopes near Tucson, Arizona.
Left Gene and Carolyn Shoemaker at the telescope with which comet SL9 was discovered.

Spacewatch

The search for asteroids was revolutionized when electronic detectors replaced photographic film.

These detectors can register more than 70 percent of all arriving photons, or particles of light, meaning that observations can be made of far fainter and more distant targets. An additional advantage is that the images can be analyzed by computer, avoiding the eye-straining task of scouring photographs for the tiny streaks that might indicate an asteroid.

The trouble is that these electronic detectors can be manufactured only up to a certain size,

typically a couple of inches (5 centimeters) on a side, whereas the image area of a large Schmidt telescope might be a foot (30 centimeters) wide. So the large photographic plates used in such systems would need to be replaced by a vast array of electronic eyes, which gives rise to a number of technical problems.

But there are ways to use these detectors in the search for asteroids. A conventional telescope might cover only a small area of space at any given instant, but by sweeping it across

the sky, a large area may be scanned, as long as the readout is taken at the same rate as the scan is occurring. Tom Gehrels, a professor of planetary science at the University of Arizona, realized this in the early 1980s, and so began the Spacewatch project.

Old Telescope, Modern Detector

Amazingly, the telescope system Gehrels uses for Spacewatch is the oldest instrument at the Kitt Peak observatory near Tucson. Built in the 1920s, it is already an antique, but it delivers results that its makers could not have dreamed of.

It was partly the age of the telescope that drove Gehrels to his solution. Over the years its drive mechanism, which should keep it pointing at the same part of the sky — even as the Earth turns — had worn so badly that the movement was too unsteady to be suitable for normal observations. Instead, in the Spacewatch scheme, the telescope would be held stationary, letting the Earth's rotation cause it to scan a strip half a degree wide across the sky. In an observation lasting twenty minutes, the strip is 5 degrees long. By repeating the scan twice, three long images are produced, and these can then be compared to find objects that have moved between times.

Spacewatch began searching for near-Earth

Above The new 72-inch (1.8 meter) aperture Spacewatch telescope.
Top left Tom Gehrels, founder of the Spacewatch project.
Left The Kitt Peak observatory in Arizona, where the idea for the Spacewatch project first took root.

asteroids in late 1989, and it was immediately successful. The project has made twenty or thirty discoveries a year ever since. Many of these are large Earth-crossers, typically half a mile across. In addition, Spacewatch has made visible a whole population of previously unknown objects.

Finding the Babies

The limited sensitivity of photographic emulsions meant that previously discovered near-Earth asteroids were all relatively large: the smallest was more than 300 yards or meters across.

You might imagine that a small target nearby would be as easy to detect as a larger one that was more distant, but this is not the case. The problem is that asteroids relatively close to the Earth have large angular motions. A large Earth-approaching asteroid would typically be discovered when it is half an AU away, and crossing the sky at about 1 degree per day. On a photographic plate it would register as a faint, short line. If it were ten times closer it would be brighter, but the line would also be longer because its apparent motion would be 10 degrees per day. A large asteroid might be detected when it was this close, but a smaller one would pass unnoticed.

For this reason, no small asteroids were known when Spacewatch began operations: there was a yawning gap in our knowledge concerning objects up to a few hundred yards in size. The Spacewatch team soon began to spot asteroids in this range, finding some as small as 5 or 10 yards across, and many others

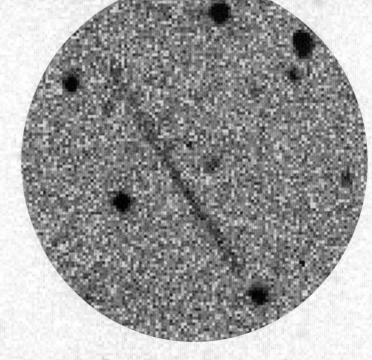

Background and above The individual pixels are obvious in this electronic image collected by the Spacewatch team. The streak above was produced by a small asteroid passing close by the Earth, seeming to zip between the stars.
Below The domes of the two Spacewatch telescopes at Kitt Peak.

as big as a house, a school, or a city block.

These small bodies were themselves interesting, but they also opened people's eyes in another way – they were very close to home. Several of these small asteroids were picked up as they passed closer to the Earth than the Moon. It was no longer possible to believe that asteroids keep their distance.

Bigger Plans

The initial Spacewatch campaign led the way in the electronic and automated discovery of asteroids, but the ancient telescope employed still has an aperture of only 36 inches. To go fainter still – and pick up small asteroids when they are much farther away – a larger telescope is needed, preferably something more modern.

Spacewatch II is a new telescope to be used in the search for Earth-threatening objects. Not only does it have a 72-inch aperture, it also has a mounting that allows it to scan quickly across the sky, enabling its operators to cover a much larger area of the sky each night, and to detect asteroids before they get too close for comfort.

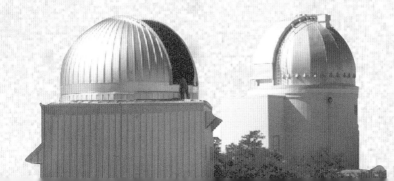

Competition and Collaboration

● **Researchers pursuing the same goals share their information. But the essence of scientific progress involves friendly rivalry as each team tries to be the first to get a result.**

In looking for near-Earth asteroids, the discovery rate you can achieve depends upon the volume of space that can be searched. This leads to the idea of a search cone: your telescope is at its apex, and the field of view of the detector defines its cross-sectional area. If you have a square detector, then the cone is shaped like a pyramid standing on its point. The height of that pyramid depends upon how faint an object you can detect. With a large aperture telescope and a high-efficiency detector, you can peer fairly deep into space. The Shoemaker-Levy team used a smaller telescope with a low-efficiency detector (photographic film), but they had a wide area, producing a short but fat pyramid. The Spacewatch project uses a larger telescope with a better detector, but the area covered is smaller, forming an elongated, narrow pyramid. The volumes of those two pyramids are similar, so their discovery rates are comparable.

Looking Up from Down Under

In Australia, we realized that the large U.K. Schmidt Telescope, operated as part of the Anglo-Australian Observatory, should be registering many near-Earth asteroids on the photographic plates obtained as part of a survey of the southern sky. Because this Schmidt camera is big, with an aperture of 48 inches (1.2 meters), the search pyramid would be deep. Since it covers an area of the sky 6 degrees on

a side, the volume searched is quite substantial. On the other hand, only a handful of photographs are obtained in any night. These photos have long exposures, making the asteroids produce streaks that might be recognized if they were scoured using a microscope. This needed to be done within 48 hours of the exposure; otherwise the asteroid would have moved on too far for it to be picked up again using another telescope at the Siding

Above The Klet Observatory in the Czech Republic is one of the most prolific trackers of NEOs. This photograph was obtained by having the camera shutter open for several minutes making the stars into streaks and rotating the telescope dome with an inside light on, so that the whole of the interior can be seen. **Right** Rob McNaught outside one of the smaller telescopes at the Siding Spring Observatory in Australia.

Spring Observatory on subsequent nights.

With my colleague Rob McNaught, a search program was started in 1990, and it was very successful, discovering dozens of Earth-approachers over the following six years.

Perhaps our most important contribution, though, was in tracking asteroids discovered elsewhere. The search teams in the United States would often find objects moving south, beyond their reach. Alternatively, an asteroid found in the northern sky might be due to reappear in the south a couple of years later, on its return from the main belt region. It was essential that these asteroids were picked up, otherwise they would soon be lost. That would be like finding a needle in the cosmic haystack, and then allowing it to fall back in among the hay.

Around the Globe, Around the Clock

Any true global asteroid tracking program that deserves the name requires extensive international cooperation, pooling the work of researchers based in observatories spread far and wide. Although most of the search activity occurs in the United States, brilliant efforts are also conducted elsewhere. Asteroids are tracked and their positions are relayed to the central node of the International Astronomical Union, of which we will learn more a little later.

Professional observatories in Canada, France, Italy, Japan, the Czech Republic, Chile, and elsewhere have been involved in such work. But

Above Amateur astronomer Gordon Garradd with his telescope in the Outback of Australia. Garradd has made invaluable measurements of NEO positions, and provided a number of photographs for this book.

much of the effort comes from enthusiastic amateurs.

It might seem strange to think that someone using a backyard telescope could contribute to a project as high-tech as tracking Earth-threatening asteroids, but this is precisely what happens. The rapid development of sophisticated electronic detectors soon made these systems available at fairly modest prices – a few hundred dollars for the simplest – in forms suitable for strapping onto the back of telescopes owned by many amateur astronomy enthusiasts. With apertures ranging from 6 to 10 inches (15 to 25 centimeters), these are not the small telescopes you might see in a department store, but at the same time they are still fairly modest in cost, and available to anyone who takes a hobby seriously. Interfaced with a personal computer, such a system is as powerful as the best telescopes that were available to professionals not so long ago, when photography reigned supreme.

Properly equipped amateur astronomers are therefore making an essential contribution to the effort to track near-Earth asteroids. No sooner is a discovery announced on the Internet than a posse of enthusiasts around the globe sets off in hot pursuit, more often than not supplying accurate measurements of its trajectory within hours or days of the announcement. Or, in some embarrassing cases, showing that the apparent asteroid does not exist, but is merely a false reading produced by the search software of the professional "discoverers."

NEAT Work

The next major search program to use electronic detectors was Near-Earth Asteroid Tracking (NEAT). The NEAT team – based at the Jet Propulsion Laboratory in Pasadena, California, and directed by Eleanor Helin – has made many significant discoveries of asteroids and comets.

Helin began her astronomical work in the company of Gene Shoemaker. The pair collaborated when they started using the small Schmidt telescope at Palomar Observatory in the early 1970s to look for Earth-approaching asteroids. But the work of the Spacewatch team showed that the days of photographic searching were coming to an end.

The Heights of Hawaii

Scientists are nothing if not adaptable: they have to be, because the technology changes so rapidly. Helin and her team spotted an opportunity to get into the automated electronic search game very quickly.

The U.S. Air Force operates several wide-field cameras for satellite surveillance. These form a network called the Ground-based Electro-Optical Deep-Space Surveillance (GEODSS)

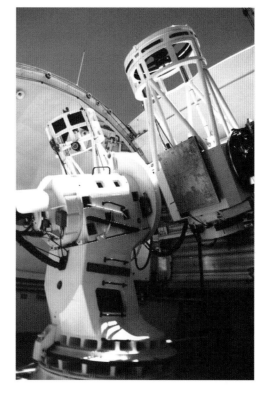

Opposite An Earth-crossing asteroid found by the NEAT team betrays its presence by moving relative to the background stars between three frames exposed about thirty minutes apart.
Left Telescopes at the Haleakala observatory in Hawaii used in asteroid tracking.
Bottom left Eleanor Helin, director of the NASA side of the NEAT project, which also involves U.S. Air Force personnel.
Below An aerial view of the Haleakala observatory. At far right, protruding from the roof of its protective building, is the large telescope shown opposite.

system. To the military, "deep space" implies distances out to the geostationary band of communications satellites, about 22,400 miles (36,000 kilometers) above the equator. To an astronomer, that's just our cosmic backyard. In fact, it's barely beyond the doorstep. We need to look much deeper for asteroids and comets.

To monitor satellites, the GEODSS usefulness is largely limited to the hours close to dusk and dawn, when the sky is dark but the satellites are still illuminated by the Sun. Looking overhead at midnight, satellites are in the terrestrial shadow so they cannot be tracked. But that is just the time – the several hours of pitch darkness before and after midnight – that astronomers are best able to search for asteroids. This led to the idea of a complementary program, tracking satellites at the start and end of the night, and searching for asteroids during the hours in between.

One of the GEODSS systems is located on the heights of the island of Maui in Hawaii. Helin conceived the plan of collecting data during the night and then sending them on to JPL, where her team could identify suspicious moving objects: asteroids and comets, rather than artificial satellites. Since it began in the mid-1990s that new project – NEAT – has been very successful, delivering many important discoveries.

Right The new 140-inch (3.6-meter) advanced concepts telescope at the U.S. Air Force observatory at Haleakala on the island of Maui in Hawaii.

Quantum Leap

A by-product of the Star Wars program was a quantum leap in the rate of asteroid discovery.

Below Grant Stokes, who leads the LINEAR project, the present leader in the NEO-discovery stakes.

As part of the Strategic Defense Initiative (often called Star Wars), the U.S. Department of Defense spent tens of millions of dollars in developing an electronic detector system that downloaded its data rapidly into computer memory. This innovation got around one of the major drawbacks of previous systems, and meant that a new image could be exposed while the previous image was being stored, allowing for essentially continuous data collection.

At the forefront of this work were the Lincoln Laboratories of the Massachusetts Institute of Technology. At that institution, others were involved in satellite tracking and the like using an observatory complex near Socorro, New Mexico. That site included not only operational GEODSS cameras, but also two identical systems that were rarely used. The leader of the Lincoln Labs team, Grant Stokes, arranged for a series of test observations to be made with one of those cameras, starting in 1996 using a conventional detector system. As expected, their results paralleled those of the NEAT team, who had by then started their work in Maui. By the middle of 1997, a handful of new NEOs had been discovered from the New Mexico site, along with dozens of other asteroids. Stokes reckoned that it would be a simple matter to scale up to the larger, more efficient detector that the military had developed. With that plan in place, he reasoned, their NEO discovery rate should rocket. And so the Lincoln Near-Earth Asteroid Research (LINEAR) project was born.

LINEAR On-Line

Since the spring of 1998, the LINEAR project has been routinely scanning the night sky for NEOs. Using a detector containing around five million pixels, it is able to cover an area about ten times as great as that of the Moon, with no time lost for data readout. By covering each celestial patch three times, moving objects are

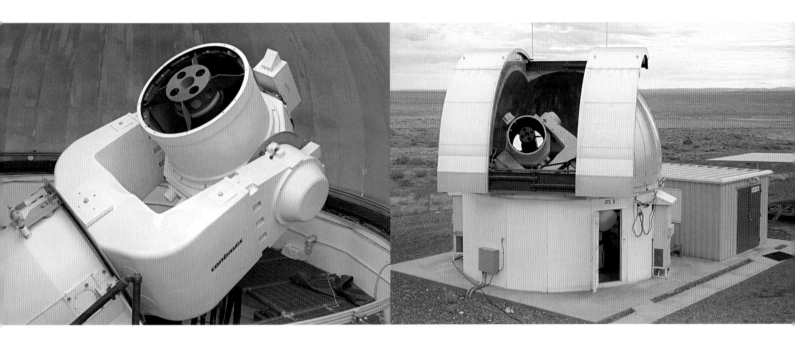

identified and additional observations can then be made. Data collection is mostly automated and the second camera system is now in use, doubling the sky coverage.

LINEAR's output has been phenomenal. In the first two years of operation, almost three million asteroid observations have been supplied to the central data repository. Four hundred thousand asteroids were detected; 60,000 of these were recorded in sufficient detail to be given new designations (such as 2000 JG5).

No doubt some of these will later turn out to be already-known main belt asteroids, but that is of little concern. The significant point is that the LINEAR team has discovered more than 400 NEOs. Previously, the global discovery rate was peaking at between five and ten per month. Now LINEAR alone is finding Earth-approachers at more than one per night.

Have the other programs become irrelevant? Not at all. The NEAT team now has a similar detector system on the GEODSS camera on Maui, and they are using other telescopes at the same site for NEO observations. And the Spacewatch team is testing out its new 1.8-meter telescope at Kitt Peak. Although this system covers a smaller area of sky than the GEODSS cameras used by LINEAR and NEAT, its much larger aperture enables the Spacewatch team to go deeper into space and see ever fainter asteroids. The next phase of competition between these NEO search teams is just around the corner.

Above left and right One of the automated 1-meter (40-inch) aperture cameras used by the LINEAR team to search for asteroids and comets.
Below The observatory site near Socorro, New Mexico.

Other Observers

● **Although the Spacewatch, NEAT, and LINEAR projects provide the majority of NEO discoveries now being made, there are other important contributors.**

Right The 72-inch (1.8-meter) aperture Plaskett Telescope operated by the National Research Council of Canada is used for astronomical projects such as the tracking of asteroids and comets.

Some of these other groups have turned up especially interesting objects from time to time. One example is 1996 JA1, which quickly became the subject of newspaper headlines.

Arizona Efforts

Asteroid 1996 JA1 captured the public's imagination because it came so very close to the Earth – at 280,000 miles (450,000 kilometers) only just farther away than the Moon. An estimated half mile (1 kilometer) in size, it is the largest asteroid spotted to date at such a near-miss distance. It was found by another University of Arizona project, the Catalina Sky Survey, directed by Steve Larson.

While the Spacewatch project covers a relatively small area (half a degree wide, the apparent width of the Moon), it can detect very faint asteroids. The LINEAR and NEAT projects

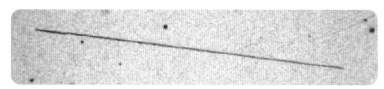

Above A photographic trail produced by asteroid 1996 JA1 as it shot by the Earth, not missing us by much.
Right A false-color view of Comet Machholz 2, which was observed to release several fragments in the mid-1990s.

cover a wider section of the sky, about a degree and a half on a side and eight or nine times as large an area at any instant, but they need slightly brighter asteroids for successful detection.

The Catalina project takes this trend even further: a smaller telescope covers a huge swath of the sky – 8 degrees on a side – but picks up only the brighter asteroids. Their system is similar to the Schmidt camera at Palomar employed by the Shoemaker and Helin teams, but adapted for electronic detectors (1996 JA1 was discovered on photographic films), and with the aperture widened to admit more light.

Another group in Arizona has followed a similar line. Rather than basing the project in Tucson, in the arid south, they work at the Lowell Observatory in Flagstaff. There, among the pine trees, where the Grand Canyon cuts through northern Arizona's high plateau, the LONEOS (Lowell Observatory Near-Earth Object Survey) project is run by Ted Bowell and his team. Like their counterparts at Catalina, the LONEOS researchers have adapted an old Schmidt camera with an extensive field of view, commencing routine operations in early 2000 after a couple of years of preliminary sky scanning. A substantial number of interesting NEOs has already been identified.

Spaceguard Canada and Japan

The largest telescope routinely used for obtaining positions of NEOs is not located in any of the places one would expect to find a major observatory, but rather in a very rainy part of Canada. At the University of Victoria, in British Columbia, Jeremy Tatum and David Balam provide an invaluable service to the NEO seekers south of the border, tracking the progress of asteroids soon after the U.S. search teams have discovered them. If it weren't for this follow-up, many would quickly be lost.

In Japan, too, a number of observers track newly spotted asteroids. Trans-Pacific asteroid search activity has recently been enhanced by the development of a telescope system dedicated to tracking man-made items in Earth's orbit, as well as the asteroids and comets that are the targets of the Spaceguard project.

France, Germany, and Italy

Three European nations that have been closely involved in the work of the Spaceguard project are France, Germany, and Italy. In France, Alain Maury and colleagues working on the Schmidt telescope at l'Observatoire de la Côte D'Azur have found several NEOs, such as the famous asteroid Toutatis, named for a Gallic god (by whom many sacred oaths are sworn in the French comic strip *Asterix*). While the telescopes are located at Caussols in the Maritime Alps, theoreticians working on the dynamics of asteroids are based at the associated Nice Observatory, on a hill overlooking the Mediterranean port city.

During the latter part of the 1990s, scientists at the German Space Agency in Berlin were

involved with the French team, working toward a joint asteroid search program. They helped to fit the Schmidt telescope at Caussols with an electronic detector for asteroid searching.

In Italy, too, several small search-and-tracking projects have begun, using a variety of telescope systems. Rome is the headquarters of the international Spaceguard Foundation, and home to a major concentration of dynamicists working on predictions of asteroid orbits – and hence possible impacts on our planet.

Above This map shows the locations of some of the observatories used to track asteroids. **Below** Looking like a giant milk bottle, the Schmidt telescope at Caussols, France has turned up several important near-Earth asteroids.

The Celestial Mechanics

Analyzing the data collected by other astronomers is a full-time occupation for some NEO researchers.

Many people imagine that astronomers spend their lives gazing through telescopes at the heavens. In fact, even observational astronomers tend to devote much of their time to other tasks, in particular analyzing the data they collect. A single night of observation may provide data that take a month or more of effort to understand.

But some astronomers never go near a telescope. They simply make use of the information gathered by others. I wrote "simply," but this part of the research is often much more complex and daunting than the collection of the actual data. (On the other hand, at least they don't have to stay up all night!)

The theoreticians who study the movement of NEOs have specialized knowledge of what is called celestial mechanics: the dynamics of objects in space. The term celestial mechanic conjures up an image of a lab-coated individual using a wrench to fix a malfunctioning star, and so they are more commonly called dynamicists. Whichever term you prefer, their science is tremendously complicated.

The Minor Planet Center

The Minor Planet Center, under the direction of Brian Marsden, is located at the Harvard-Smithsonian Center for Astrophysics, in Cambridge, Massachusetts. It is approved by the International Astronomical Union as the appropriate central repository for observed positions of all asteroids and comets; as a result, Marsden and his small team handle a continuous stream of incoming data from observers spread around the globe.

On any day many thousands of individual observations may arrive over the Internet. These are checked, largely automatically, for consistency with known asteroids, or categorized as new discoveries. The heavens are teeming with rocks, all flying in different directions but in predictable ways. In consequence it is possible to write software to identify errors in the observations, or to look for previous observations of an object stored in the data banks. With millions of observations stashed away, it is often the

Top left Brian Marsden, director of the Minor Planet Center, is the ultimate arbiter on matters concerning asteroids and comets. **Above** Paul Chodas (left) and Don Yeomans (right), dynamicists at NASA's NEO office, provide advice on whether specific asteroids and comets pose a threat to us.

ARE WE SITTING DUCKS?

Right Will this asteroid hit our planet, or will it slip silently by? This is the sort of question that the celestial mechanics must answer if we are to identify our potential nemesis rock.

Opposite left and far left A small asteroid, heading for the ocean, enters the atmosphere. It is to identify and track objects like this that constant vigilance is necessary.

case that an asteroid spotted in the sky now was also detected several years ago, but with insufficient data having been collected in order to define accurately its orbit.

NASA's NEO Office

The Jet Propulsion Laboratory in California has a team of dynamicists comprising NASA's NEO Office, led by Don Yeomans. Announcements made by NASA headquarters in Washington on the subject of hazardous asteroids are checked beforehand by the experts at JPL. This office also correlates the various asteroid and comet projects with which NASA is directly involved. Others at JPL working on NEO observations include Helin's NEAT team, and also the radar group whose results will be discussed shortly.

The City of the Leaning Tower

There are several institutions and individuals elsewhere who concentrate on theoretical studies of NEO dynamics. Russia has a well-regarded reputation in this area, particularly the Institute of Applied Astronomy in

St. Petersburg. At Armagh Observatory in Northern Ireland a small group under Mark Bailey pays particular attention to comets with intermediate orbits, like Comet Halley. Nice Observatory has already been mentioned.

In recent years, however, many of the announcements concerning asteroids that could potentially evolve – within a few decades – into Earth-impactors have come from the group at the University of Pisa, led by Andrea Milani. This team, with various collaborators in Italy and elsewhere, has pioneered techniques for making identifications of avenues through which specific newly found asteroids might perhaps come back to hit the Earth in a few orbits' time. This is important because it directs the attention of observers to the objects that are potentially most hazardous. This, in turn, leads to more measurements being obtained, which allows their orbits to be better defined. Up until now, thankfully, the outcome has always been the same: the improved determinations quickly show that the objects' future paths do not intersect the Earth. But that may not always be the case.

The Riches of Radar

We have discussed using optical telescopes to search for asteroids and comets; radio telescopes have not been mentioned. Surely this other fundamental tool of astronomy must have a role to play?

Radio telescopes are used in some forms of comet observation: for example, in determining the types of molecules that exist in the coma and tail. But comets are not strong radio sources. Neither are asteroids. Radio astronomers mostly study distant stars and galaxies.

With optical telescopes we detect asteroids and comets through the sunlight they reflect. But the Sun is a weak source of short-wavelength radio emission. We can generate our own radio waves, though, and get echoes from asteroids and comets that way. This is the principle of radar.

Planetary Radars

An airport radar may need to monitor airplanes out to ranges of a few hundred miles. That's a tiny distance on the scale of the solar system: astronomers might want to bounce radar pulses off an asteroid a hundred million miles or kilometers away, which would require an extremely powerful radar system.

But there are other complications. If an airplane is a hundred miles or kilometers away, it takes just a millisecond or less for the radio pulse to reach it and then return to the radar site. The round-trip time for a radio pulse, traveling at the speed of light, to reach and return from an asteroid a hundred million miles distant is about eighteen minutes. So rather than getting an essentially instantaneous echo, you need to have your receiver equipment ready to pick up the return at precisely the right moment.

Apart from that, the echo you get is shifted in frequency. The Doppler effect, which explains

why the note of an ambulance siren changes as it whizzes past you, applies to light and radio waves as well. The speed of the target object causes the returned signal to be shifted in frequency. In the case of a car or an airplane, the shift is small but measurable. That's how a police radar speed trap works.

An asteroid in space, though, is traveling typically at one part in 10,000 of the speed of light. This has the effect of shifting the returned frequency outside of the bandwidth of the signal originally transmitted. On the one hand, this means that the receiver needs to be tuned to the expected frequency of the returning echo;

Left The 230-foot (70-meter) radio telescope at Goldstone in California is part of NASA's Deep Space Network. It has been used to bounce radar signals off near-Earth asteroids. **Below** Occasionally, the network of antennas forming the Very Large Array in New Mexico has been employed to receive the returned echoes from asteroids.

ARE WE SITTING DUCKS?

Defining Orbits

The complexity caused by the combination of long delay in receiving the echo, the Doppler shift, the weakness of the echo, and the narrowness of the beam produced by a large dish means that radars are simply no use in searching for unknown NEOs. Even if these problems could be overcome, there is another insurmountable drawback. Interplanetary space is populated by a huge number of smaller meteoroids, and the diffuse echoes from them would drown out the signals from the larger objects we want to find. It would be like looking for a snowman in a snowstorm.

Radars are very useful, though, for defining orbits accurately once the initial parameters are known. With an optical telescope one measures positions of the target, relative to known stars, in the sky plane: the plane at right angles to your viewing direction. The uncertainty in the positions may amount to some hundreds of miles or kilometers. Radar, however, renders a measure of the distance away in the viewing direction with higher precision, down to yards or meters. Also, the speed along that line is determined (from the Doppler shift), to a matter of inches per second. In terms of getting an accurate orbit, one radar detection may be worth as much as ten years of optical tracking. If we are to be able to predict an impact well ahead of time, radar observations are invaluable.

One note of caution: there is no planetary radar in the southern hemisphere. We'd be in big trouble if a potential Earth impactor appeared in the southern sky, as we'd be unable to refine our orbit determination using radar. Our soft underbelly is completely exposed.

Above The radar dish at Evpatoria in the Crimea, used in conjunction with a German radio telescope to study asteroids.
Top The Arecibo radar in Puerto Rico has a dish 1,000 feet (300 meters) across. It is the most powerful radar used for observations of asteroids and comets, but it is not steerable.
Below The shape of the Earth-approaching asteroid named Bacchus has been deduced from radar data, despite the fact that it's just a pinprick of light in an optical telescope.

on the other hand, a measurement of the Doppler shift allows the line-of-sight speed to be determined, and so improves our knowledge of the asteroid orbit.

In order to be able to detect radio echoes from distant solar system objects, you need a powerful transmitter that delivers megawatts of radio energy, and a very large dish to provide a narrow beam that can pinpoint the target.

One such system is the Arecibo radar in Puerto Rico. Its 1,000-foot dish is fixed, but some steering is possible by moving around the horn that collects the returned echoes. A recent upgrade has made this the most sensitive planetary radar system on Earth.

The Goldstone planetary radar system in California has various dishes ranging from 110 feet to 230 feet in size, and can be used for transmission and reception. The transmitter at Goldstone has also been used in conjunction with the Very Large Array (VLA) in New Mexico in radar studies of solar system objects.

The only other planetary radar in existence is a binational system: a large dish at Evpatoria in the Crimea is used as a transmitter, and the returned echoes are received by another radio telescope located in Bonn, Germany.

Rock, Iron, and Ice

Radar echoes from asteroids can tell us much about their compositions and sizes.

Astronomers using optical telescopes have found that asteroids have various spectral types. In other words, their colors are different. This is not unexpected – we see many distinct types of meteorites.

Similarly, the overall fraction of sunlight reflected – called the albedo – varies from asteroid to asteroid. Some have albedos as low as a few percent, while others scatter as much as half of the light that hits them. For comparison, the mean albedo for the Moon is about 11 percent, for Mars it is 15 percent, while for the Earth it averages out to nearly 37 percent, as a result of the snow-covered poles, cloud cover, and the relatively bright desert areas.

Because asteroids are so tiny, it is generally not possible to measure their sizes directly. A few of the larger ones have been resolved using the Hubble Space Telescope, but most remain pinpricks of light in a telescope. By timing how long a star disappears from view as an asteroid flies across it, an infrequent but predictable event called an occultation, a handful of main belt asteroids have been measured. A few others have been visited by space probes. But most have not had any kind of direct scrutiny.

It is possible to infer an asteroid's size using ground-based telescopes, however. In the visible region of the spectrum, the light one detects is the sunlight that is reflected. The rest of the sunlight striking any solar system body is absorbed, heating it. That absorbed energy is reemitted as infrared radiation. By comparing the apparent brightness of an asteroid in the visible and infrared parts of the spectrum, astronomers are able to work out how big it must be, to within a reasonable degree of certainty. In

general, the sizes and albedos are determined in this way.

To get the colors, astronomers use different filters on a telescope, splitting the visible spectrum into zones. The relative brightnesses are then used to allocate asteroids into discrete classes, depending upon both their colors and albedos: are they dark and gray, or brighter and bluish? It is common to talk about asteroids as being D-types, S-types, M-types, and so on, with more than a dozen classes being employed. Many of these can be associated with distinct types of meteorite found on Earth. S-types get their label because they are like stony meteorites. M-types are like metallic meteorites.

Spinning Asteroids

Asteroid sizes may also be determined through the use of radar. The strength of a radar echo depends to some extent upon the size of the object: all other things being equal, an asteroid that is twice the size of another will give an echo four times as strong, because its cross-sectional area is four times as big.

But things are not quite that simple. The scattering properties of any object depend upon the texture of its surface and the wavelength of the radiation reaching it. In the case of visible light, a polished table looks different from one covered in dust or sand. The radar systems used in observing asteroids have typical wavelengths of an inch or two. In consequence, a flat rocky surface will give a markedly different echo from one made up of a pebbly texture, which is what one might anticipate.

To complicate things further, asteroids spin. This means that the surfaces scattering radio

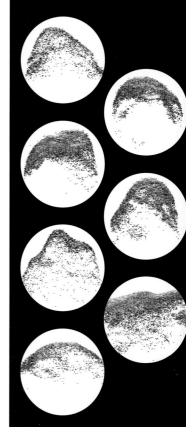

Above Crude radar images of the 2-mile (3.5-kilometer) asteroid 1999 JM8 indicate only its basic shape, but even these pictures show that it has a rounded, cratered profile.
Above right Radar images of the Earth-approaching asteroid 6489 Golevka, which is about half a mile (800 meters) in size. The color coding represents the surface texture and slopes of this rocky object, discovered in 1991a.

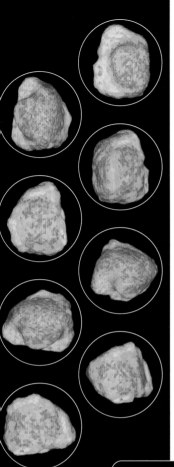

waves back to the home receiver will be changing all the time, albeit slowly, because a typical spin period ranges from a few hours to a day or so. Compared to the motion of the center of mass, one side will be moving toward the Earth at any instant, while the opposite side will be moving away. The result is that there are different Doppler shifts for the echoes from different parts of the target. Those tell us how fast it is spinning, and eventually it is feasible to map out the size and shape of the asteroid.

Anomalous Returns

The echo strength obtained depends not only on the surface texture of an asteroid but also on its composition. Returning to the shiny tabletop analogy, we are familiar with the different reflections produced by tables made of glass, polished metal, or wood. Materials that are electrical conductors scatter the incoming waves better than insulators, such as rock. In this way it has been possible to use planetary radars to

show there is subsurface permafrost on Mars.

Some asteroids show anomalous echoes that are very strong in relation to the size of the objects that produce them. These are most often M-type – metallic – asteroids. We think they are made of nickel-iron alloys similar to many meteorites that we find on Earth.

When investigating the properties of a particular asteroid that may strike the Earth, it is obviously helpful to know its composition. A loose aggregate of boulders with a low density represents quite a different level of risk than a lump of solid iron with a density five times as high. Radar observations are invaluable in making such determinations.

Below Color-coded radar echoes show how the Earth-crossing asteroid 4769 Castalia rotated during the observation sequence. This enabled researchers to construct a computer model of its shape, with two distinct lobes apparent in the images at left. Castalia is about 1.5 miles (2.4 kilometers) long.

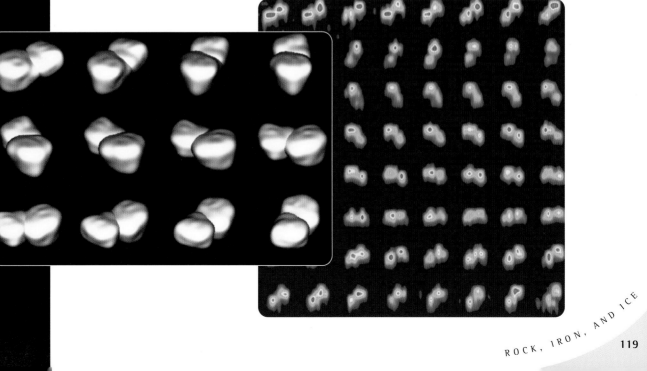

The Asteroidal Zoo

Astronomers often talk about the cosmic zoo: how their telescopes deliver images of bizarre phenomena that populate the far reaches of the universe.

But there's a major menagerie closer at hand. We've seen some pretty peculiar sights in the solar system: a volcano higher than Mount Everest on Mars, sulfur volcanoes on Io, and ice eruptions on Triton. But nothing compares with the shapes of asteroids revealed by radar observations.

A spinning asteroid produces a range of radio frequencies in its echoes because some parts are moving faster than others. Some surfaces are closer to us at any given instant while others are farther away, and this results in a tiny but measurable spread in the times at which the echoes are received. These combined effects enable us to map out the shape of an asteroid using what is termed the delay-Doppler technique. In this way we have obtained profiles – to unprecedented levels of detail – of asteroids that were hitherto nothing more than moving points of light.

Cosmic Peanuts

Castalia and Toutatis are two large near-Earth asteroids whose strange shapes have been unveiled by radar. The former appears to have two main lobes, which may represent a pair of individual rocks held together by mutual gravitational pull. Composite images show a narrow neck between them, but there is insufficient resolution to know whether the two are simply in contact at that point, or whether the junction is filled with a mass of other rocky material. Castalia looks like a very large peanut.

Toutatis, on the other hand, has all the appearance of a misshapen potato. This raises the question of whether most asteroids are single pieces of rock, groups of a few large

lumps, or rubble piles – accumulations of boulders, held together by gravity until knocked apart by some disruptive force.

Until quite recently the rubble pile theory held sway, because no asteroid had ever been observed to spin so fast that self-gravity would be inadequate to hold it together. Now, though, we know of several rapidly rotating asteroids, that the laws of mechanics indicate must be single rocks.

But nothing with regard to asteroid rotation is ever simple. A sphere like a planet will spin on a single axis, but odd-shaped bodies are prone to have more complicated rotatory behavior, with multiple axes of rotation. Toutatis is a particularly good example. In many ways, it makes more sense to think of it tumbling along its orbit, rather than spinning.

Dog Bone in Space

One final specimen in the asteroidal zoo is 216 Kleopatra. Discovered in 1880, this main belt asteroid gave no hint that it was anything out of the ordinary until the upgraded Arecibo obtained radar echoes to profile its shape. What resulted was something of a surprise.

Kleopatra is shaped like a giant dog bone, 135 miles long and 58 miles wide (217 by 93 kilometers). Just how it attained such a perplexing form is a mystery. Perhaps it was sculpted by some phenomenal interasteroid collision. Perhaps it accumulated this way through some process we have yet to understand. Of course, those who believed that the "face on Mars" was evidence of alien visitations will have their own ideas.

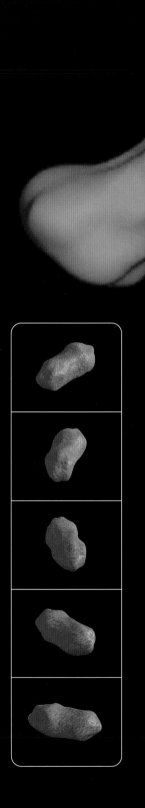

ARE WE SITTING DUCKS?

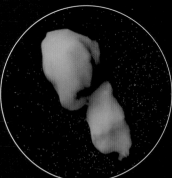

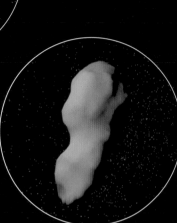

Right Like many other small objects, Toutatis spins not just around one, but around several distinct axes. Overall, it appears to tumble along through space.

Sequence far left The cosmic dog bone. Radar images of asteroid Kleopatra show that it is shaped like a dog's bone, or maybe a dumbbell.

Above left A close-up of the huge main belt asteroid Kleopatra.

Sequence left Various aspects of the large Earth-crossing asteroid 2063 Bacchus, obtained from radar data.

Sequence left Asteroid 4179 Toutatis in various stages of its rotation. This object is about 2.5 miles (4 kilometers) long and frequently passes close by the Earth.

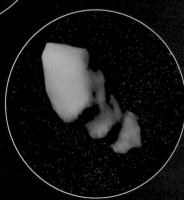

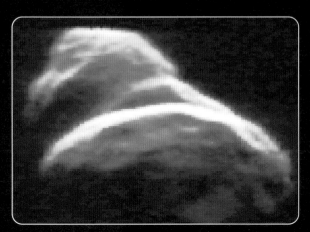

Left An end-on radar view of Toutatis. By collecting many more echoes, it is possible to smooth out the blockiness of the data and obtain a more accurate picture of an asteroid's shape.

Let's Talk Money

How much should we spend on tackling the impact hazard?

With an estimate of how many of these objects share the inner solar system with us, we can work out roughly how often a cataclysmic impact might be expected. By looking at the craters scattered over the Earth's surface, and seen on other planets, we can work out how much energy is released in these major collisions.

Whatever way you look at it, a major impact is a horrific event to imagine. Although we must be careful when we extrapolate from less energetic explosions, it is quite straightforward to scale up from the largest nuclear detonations and see how big an asteroid or comet impact would need to be to cause a global upset. We are talking about an event that would kill at least a quarter to a half of mankind.

Most estimates indicate that an impact releasing energy equivalent to between 100,000 and one million megatons of TNT would be sufficient to cause such a worldwide catastrophe, no matter where the impact occurred. (For comparison, the most powerful weapon ever tested was a 65-megaton hydrogen bomb.) The size of asteroid required to release such energy is quite small: according to our calculations, it need only be between about 0.7 and 1.5 miles (1 to 2 kilometers) in size.

How often do such events occur? Again, we have to rely on our best estimates, but between once every 100,000 years and once every 500,000 years is about right. Adding in comet impacts will reduce the timescale.

Using this information, we can show that your chance of dying due to an asteroid impact is about one in 10,000. The precise probability may

be a little lower, or a little higher, but we think this is a reasonably accurate figure given the state of our knowledge at present.

Comparative Risks

To get an idea of how concerned we should be by this figure, we need to compare the impact hazard against other more familiar risks we run in daily life. For example, people tend to worry about flying. Taking the last ten years of data for major aviation accidents in the United States and Europe, we can calculate that the likelihood of dying in a jetliner crash is around one in 30,000. A couple of jumbo jet crashes in a year may increase that to one in 20,000. A quiet period may lower it. But you are more likely to die in an asteroid impact than in a plane crash.

When all factors are taken into account, road

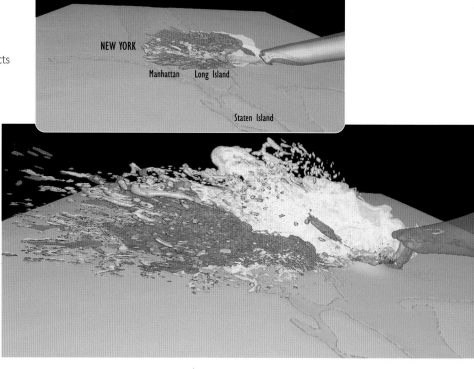

Above This computer model shows the results of an oblique impact by an 0.9 mile (1.4 kilometer) asteroid into the Atlantic Ocean just off Long Island in the United States. New York City is quickly engulfed by rock and water that are displaced by the impact (top), followed soon after by a huge mass of material that penetrates deep inland (bottom). The rain of rock and water devastates all life in the region, and global atmospheric disturbances result from the ferocity of the explosion.

crashes are the most frequent cause of accidental death, followed by house fires and electrocution. But among natural accidents, such as tornadoes, floods, snakebites, and lightning strikes, asteroid impacts come out on top.

The Economics of Armageddon

We have shown that NEO impacts pose a significant risk. How much should we spend making sure that one does not occur? If things were done according to logic, the expenditure would be proportional to the risk. This is not the case – politicians spend public money broadly in tune with voter wishes, and most people do not view asteroid defense as a priority – but it is a useful starting point.

Averaged over a long period, the mean number of deaths of U.S. citizens due to NEO impacts may be estimated at about 300 to 400 per year. (That is, two or three times higher than through airplane crashes, which we spend billions

Automobile accident	1 in 100
Homicide	1 in 300
Fire	1 in 800
Accidental shooting	1 in 2,500
Electrocution	1 in 5,000
Asteroid/comet impact	1 in 10,000
Jetliner disaster	1 in 30,000
Flood	1 in 30,000
Tornado	1 in 60,000
Snakebite, insect sting	1 in 100,000
Fireworks accident	1 in 1 million
Botulism poisoning	1 in 3 million

Above Average accidental death probabilities for U.S. residents, based on a 1994 paper by C.R. Chapman & D. Morrison (Impacts on the Earth by asteroids and comets: assessing the hazard, *Nature*, 367, 33-40) with updates by the author.
Right The surprisingly high probability of dying as a result of an impact is explained by the sheer number of fatalities such an event would cause.

to minimize.) The reality is that in almost every year zero deaths are caused by asteroids hitting the Earth, but infrequently such a calamity will occur, and at least 50 to 100 million people will die in the United States alone.

At what level does the U.S. government value its citizens' lives? The answer is at three or four million dollars each, in terms of civil and national defense, road safety, and so on. Multiplying that by the expected death rate indicates that the annual expenditure on defending against impacts should be somewhat more than a billion dollars.

The Limit of Tolerability

This calculation is similar to the actuarial sums done in assessing insurance risks. National governments have to face the sort of risks that would bankrupt an insurance company. But they have other things to ponder. What risks do their citizens consider intolerable? In that context major disasters, such as a nuclear power plant exploding, are regarded as hugely significant even if they have only a tiny probability of occurrence.

The government of the United Kingdom employs a matrix to determine which sorts of accident are tolerable and which are not. Take some arbitrary type of accident with an annual probability of one in 100,000. According to the guidelines, a disaster with that likelihood of occurrence might be tolerated if it were expected to kill fewer than a hundred people. More than that and ways would be sought to reduce the chance: a one in 100,000 event likely to kill 10,000 people is regarded as being intolerable, and expenditure to tackle it is mandatory once the risk has been recognized.

At about that annual probability level, NEO impacts are expected to kill ten million people in the United Kingdom alone, fifty million plus in the United States, and hundreds of millions elsewhere. By definition, the hazard posed by asteroids and comets is intolerable. We must act.

Type of accidental death

Typical number of people killed per accident (figures for the United States)

Car crash — 1
Flood — 10
Jetliner crash — 100
Asteroid/comet impact — 100,000,000

Project Spaceguard

By now it should be clear to most readers that asteroids and comets do indeed strike the Earth from time to time with catastrophic consequences. What can we do to protect ourselves?

Small bodies rain down into the atmosphere all the time, but the large ones we might need to worry about — those around 5 or 10 miles in size — arrive much less frequently. Geologists have found evidence of their destruction in the shape of craters scattered all around the globe, and paleontologists have unraveled evidence for mass extinction events that might be linked to calamitous impacts by large near-Earth objects (NEOs).

Smaller impactors also pack a vast punch, though, and we have seen that an asteroid less than a mile across would release sufficient energy to cause a global catastrophe leading to the deaths of a significant portion of mankind. When you do the calculations, you realize that the risk of this causing your demise is disturbingly high, at least as great as the chance of dying in a jetliner crash. Such an event would cause the downfall of civilization as we know it.

Obviously the level of risk dictates that something should be done to tackle it. But what?

Fiction and Fact

The layman is usually surprised to learn just how small an impactor must be to release, say, a million megatons of energy. Most people have little feel for how fast the Earth is traveling around the Sun, and therefore the speed at which an NEO would run into us, and the huge amount of energy involved as a result. But good science fiction writers base their novels on real science, and over the years several books have given reasonably accurate accounts of the consequences of impacts.

One of these is *Lucifer's Hammer* by Larry Niven and Jerry Pournelle. The origin of the title is obvious. Another is Arthur C. Clarke's *Rendezvous with Rama*, published in 1973. This starts with an account of a small asteroid hitting northern Italy in the year 2077, and the devastation it wreaks upon Europe. Clarke then continues:

After the initial shock, mankind reacted with a determination and a unity that no earlier age could have shown. Such a disaster, it was realized, might not occur again for a thousand years — but it might occur tomorrow. And the next time, the consequences could be even worse. Very well; there would be no next time . . .

So began Project Spaceguard.

Although Clarke's book was fictional, this segment is based on fact, and those of us working in the area decided that it was appropriate to use the name Spaceguard for the real-life project we planned. The difference is that we see no need to wait for a disastrous impact to wake people up to the danger posed by asteroids and comets.

Public Scares

After Hermes buzzed the Earth in 1937, the news media — always on the alert for a good story — had occasionally picked up on the risk of a major impact. But all remained relatively quiet until the discovery in 1989 of the asteroid now cataloged as 4581 Asclepius. In March of that year this newly found asteroid whizzed by us at a distance of only about 400,000 miles (650,000 kilometers) — not much farther from Earth than the Moon. On the cosmic scale, that's our backyard. In fact, you could say that it missed us by the equivalent of five or six hours, in terms of how long it takes to travel such a distance. And Asclepius is a good size: several hundred yards across, and capable of delivering a blow equivalent to many thousands of megatons of TNT in a collision with Earth.

This made headlines. The idea that such projectiles do not always keep their distance came close to inducing panic in some quarters, and there were major repercussions. Suddenly people wanted to know what was being done to patrol space and monitor the threat. And so the U.S. government asked NASA to look into the matter.

Above An artist's rendering of an asteroid approaching the Earth. Were this to strike home it would kill a large fraction of humanity. Could such events be predicted long enough in advance for us to protect ourselves?

Right A color-coded satellite image of the Zhamanshin crater in Kazakhstan. It is 8 miles (13 kilometers) wide and about 900,000 years old.

Far right Even with a covering of clouds, the circular form of the 7-mile (11-kilometer) Bosumtwi crater in Ghana is obvious. It is around 1.3 million years old.

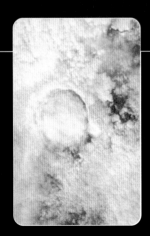

Politicians Act

● Media coverage of close misses of the Earth by asteroids prompts action.

Evidence published by the Alvarez team in 1980 that pointed to the extinction of the dinosaurs in the wake of a massive impact established the idea that asteroids and comets have struck our planet in the past, with calamitous results. We know from recent observations of asteroids with Earth-crossing orbits that the impact hazard is as great now as it was in the dinosaurs' era. Despite this, many continued to believe that such events were purely matters of historical record, with no relevance to the present day.

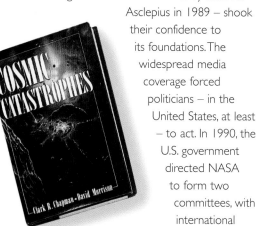

A single event – the near miss by asteroid Asclepius in 1989 – shook their confidence to its foundations. The widespread media coverage forced politicians – in the United States, at least – to act. In 1990, the U.S. government directed NASA to form two committees, with international representation, to report back on two aspects of defense against an NEO impact.

Their first brief was to establish a program of surveillance that would search for and locate any large object due to strike the Earth soon (and we'll decide what "soon" means below). Their second was to investigate how an asteroid or comet identified to be on a collision course with Earth might be deflected or destroyed.

In many ways the task is analogous to establishing a screening program for a deadly disease. Screening an entire population is expensive but essential if the disease is to be identified early enough to attempt a cure. An individual's chances of contracting the disease are small. But if screening shows that you have the disease, you need to intervene to save your life. All attempts at a cure are expensive and unpleasant, much more so than a simple screening. And they're all fraught with difficulties. But the relatively cheap first step of screening increases the likelihood that further measures will be effective.

The Spaceguard Committee

The responsibility for defining a surveillance program that would locate any NEO due to run into the Earth was delegated to a committee chaired by David Morrison of NASA-Ames Research Center in California. I was one of the twenty-four members of what has become known as the Spaceguard Committee. The name derives from the title of the NEO survey program we recommended, in homage to the fictitious program described by Arthur C. Clarke (who first conceived the idea) in his 1973 book *Rendezvous with Rama*.

A point that is often misunderstood is that any defense against an NEO impact requires a collision to be predicted many years ahead. It is not a matter of spotting our nemesis a week or so away and blasting it out of the sky. That's pure Hollywood. If we are to have a chance of defending ourselves, we need a lot of warning. Preferably decades.

This means that we need to spot NEOs while they are still a long way distant, and that requirement determines the minimum size of the search telescopes. If you are hunting for asteroids larger than about half a mile,

Above and left Clark Chapman (top) and David Morrison (above) are two of the main U.S. advocates for the Spaceguard program. As far back as 1989 they published a popular-level book, *Cosmic Catastrophes*, (far left) which explained the hazard posed by asteroids and comets.

or 1 kilometer across, and you want to detect them when they are at the far reaches of their orbits, perhaps just closer than Jupiter's path, then you need telescopes with apertures of at least 80 inches (2 meters).

The Spaceguard concept comprised a global network of at least six dedicated search telescopes, three each in the northern and southern hemispheres. The aim was to seek out all asteroids larger than one kilometer (about a half mile) in size, and to determine their orbits with sufficient precision to be able to predict any impact due within the next several decades. That is what we mean by "soon."

The total cost of the Spaceguard Survey was estimated at $300 million spread over

The Spaceguard Survey

Report of the
NASA International Near-Earth-Object Detection Workshop

David Morrison, Chair

January 25, 1992

twenty years. It has never been executed in the form described in the 1992 report. Instead, partly because no other nations agreed to join the United States in funding the project, a more modest program was agreed to, along the lines of a subsequent report presented to NASA in 1995 by a smaller committee chaired by Gene Shoemaker. This had a lower budget and a more modest aim. The present search programs in the United States largely fulfill this aim, but the smaller telescopes in use mean that the search is restricted in its range, with the result that many NEOs will remain undetected. And with no participating telescopes in the southern hemisphere, half the sky is completely exposed.

Left NASA's original Spaceguard report, published in 1992, outlined the surveillance program necessary to identify asteroids or comets on a collision course with the Earth.

Shock front produced by the comet's supersonic passage through the atmosphere

Cometary material and water ejected from the transient crater

Virtually empty column of atmosphere left in the wake of the comet

Explosion site, where the comet struck the ocean surface, forming a temporary crater

Deep ocean

Solid ocean bedrock

Left In this computer model of a comet impacting the ocean, the water is orange and the solid ocean bed is red. The atmosphere above is shaded from blue to green (blue for the low-density air of the upper atmosphere, green for the higher density air lower down). The comet arrives from upper left, leaving a blue (low-pressure) column behind it. A yellow-colored transient crater several miles or kilometers deep forms in the ocean. A huge amount of water is displaced, generating a massive tsunami.

Interception and Deflection

The other critical requirement of an NEO defense system is the ability to intercept and deflect any large impactor in good time.

If we carried out the appropriate search and tracking program and had all the NEO orbit data in hand, we would most probably discover that none is due to hit the Earth within the next century. Some might come uncomfortably close and require continued monitoring, but an impact by a large asteroid is unlikely.

This is certainly not a reason to avoid putting the plan into action, however. It's a bit like an insurance policy. You don't insure your car and then at the end of the year bemoan the fact that you've had no accidents.

But what if we did find one of these objects on course to hit us? What would we do then? This was the problem the second NASA committee was directed to consider: the tricky matter of interception and deflection.

Showdown at Los Alamos

I attended the main meeting of this committee, at Los Alamos, New Mexico in January 1992. Although that was almost a decade ago, the basics of what was discussed in that week-long session have not changed much since then. The laws of physics are unbending.

If we are to deflect or destroy an NEO heading for the Earth, the first step is to locate it. That's the job of the surveillance program. The next step is to get to it. This is much more difficult. In fact, it is rocket science.

Current spacecraft targets – the asteroids and comets to be visited in the next few years –

Left Intercepting an asteroid heading for Earth would not be a simple matter of loading nuclear weapons onto the space shuttle. To carry big payloads into deep space, larger launchers are required.

are chosen because they are comparatively easy to reach. That is, they have low speeds relative to the Earth. But the chances are that any NEO we need to intercept for the purposes of self-defense will not be one of these easy targets. We will almost certainly need a more powerful rocket system than anything currently available.

It is widely imagined that the space shuttle goes off into deep space; in fact, it never strays more than a few hundred miles above the atmosphere. To deliver some weighty device to an NEO would require a heavy lift capability. When the MIT students were working on their Project Icarus report in the late 1960s, the Apollo moon landing program was in full swing, and huge Saturn V launchers were available. That is no longer the case. If we needed to get to a threatening NEO in a hurry, we wouldn't necessarily have the rocketry to do it.

That's just one of a number of problems we face. There are others, like the level of precise guidance control required to intercept an object moving through space at tens of miles or kilometers per second. But even if we did manage to rendezvous with a potential impactor, how should we deal with it?

Simple Physics

People talk about blasting an asteroid out of the sky. That's nonsense. If you break up an NEO heading for the Earth, you simply transform a cannonball into a shotgun blast: almost all of the fragments will still hit the planet, causing nearly as much damage as would have been the case anyway. What we need to do is to give it a nudge, to push it – without breaking it up – onto a slightly different trajectory that makes it miss

the Earth. How do we do that?

It is quite easy to calculate the energy necessary to divert a 1-mile asteroid by an amount sufficient to make it miss. As a ballpark figure, a million tons of TNT would be required. That amount rules out chemical explosives, like TNT itself, because we can't fly a million tons of material of any sort to a distant asteroid or comet. Even ten tons would be difficult.

In fact, the whole idea of conventional explosives is nonsensical in this context. At a speed of just 2 miles (3 kilometers) per second, an inert mass – say a lump of lead or rock – has kinetic energy in excess of the chemical energy

of the same mass of TNT. Simply flying a huge rock into the target at 10 miles (16 kilometers) a second would have much the same effect as exploding a huge stick of dynamite, and would be simpler to accomplish.

The question comes down to one of specific energy: the energy liberated per unit mass of the interceptor spacecraft. The only feasible way to deliver enough energy is through a nuclear weapon. To divert an NEO due to hit the Earth, this is what we would have to use. As I wrote earlier about cures for a deadly disease, the possible solution is not pleasant, but it is inescapable.

Above A computer-generated depiction of an asteroid being diverted by a nuclear explosion. The weapon is detonated above the target's surface because this reduces the chance of blasting it apart. The idea is to give it a shove while keeping it in one piece.

A Job for the Military

● Should space matters be a concern of the world's defense departments?

A century ago no nation had an air force. Nowadays, command of the air is essential in many conflicts. Times have changed. Britain became a global power when, in the words of the jingoistic song "Rule Britannia," it managed to "rule the waves." But the days of the navy, if not entirely past, have at least ebbed as aerial dominance has grown in importance.

It is clear that control of space is the next frontier of military might. Not only for the purposes of security between opposing nations on the Earth, through batteries of surveillance satellites and communications links, but in the defense of all humanity against missiles from space.

If we are to have a long-term future on planet Earth, then our increasing knowledge of the projectiles that cross our Sun-centered orbit has made one thing clear: we must learn to dominate the solar system. We must contain the hazard in the same way that we have eradicated smallpox and, we hope one day, cancer and the common cold.

Oh My Darling Clementine

The modern military must have some interest in space matters. If you rely on space assets – and all branches of any defense force must do so, if only through the GPS (Global Positioning System) – then you need to know if a strong meteor shower like the Leonids is liable to knock your system off the air. The U.S. Department of Defense (DoD) already has some involvement in asteroid search and tracking through the LINEAR and NEAT programs described earlier.

Another aspect of this work is direct spacecraft investigations. After the end of the

Apollo program, it was more than twenty years before another U.S. satellite was sent into orbit around the Moon. This was the Clementine spacecraft in 1994, a testbed for various technologies developed in the Star Wars program. Clementine was built by the DoD but operated in conjunction with NASA scientists. It returned images and other data of unprecedented detail from the Moon, including information about possible ice lakes in polar lunar craters that would be an invaluable resource for future exploitation of our companion.

After being slung out of lunar orbit, Clementine was supposed to be sent on to fly past an Earth-crossing asteroid, the cigar-shaped 1620 Geographos. But a thruster malfunction set it spinning and exhausted its fuel supply, and as a result this secondary aim of the mission could not be completed.

Left The Clementine spacecraft, which mapped the Moon in unprecedented detail in 1994.

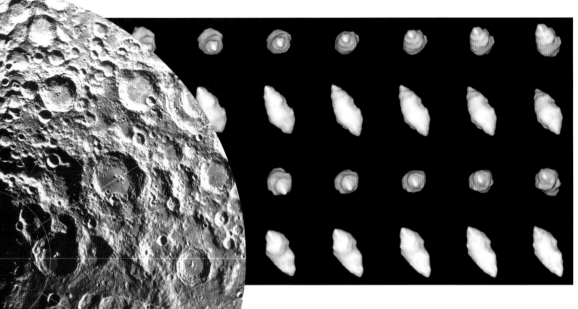

Left Radar images of asteroid 1620 Geographos, the second target of the Clementine mission, which was not reached due to a rocket malfunction.

Above The Moon's south polar region as mapped by Clementine in 1994. Ice deposits are in the perpetually shadowed crater interiors near the pole.
Left Clementine was used to measure the heights of the lunar surface and so produce this topographic map. The lowest ground is purple, grading to blue and then green, with the highest elevations in red.

A second spacecraft, Clementine II, was planned to launch in 2000. This was to be a dedicated asteroid mission. In fact, the idea was to probe under the surface of three Earth-approaching asteroids by firing harpoons about 5 feet (1.5 meters) long into their surfaces at hypervelocity, blowing craters out of them to see precisely what they were made of and how strong they were. We are still very unsure about asteroid makeup: are they mostly single rocks or are they clusters of separate boulders – rubble piles – held together by self-gravity? If we are ever going to be able to divert an asteroid on a collision course with Earth, then we must be able to intercept these objects in this way. Clementine II, however, was viewed as too aggressive by the Clinton administration, and it was vetoed in 1997.

The Deflection Dilemma

There is another aspect to all this, though. Suppose that we – mankind – develop a defense system capable of diverting asteroids away from the Earth. That sounds great, even essential to our long-term survival, if you agree with this book so far. But the downside is this: if a nation has the ability to divert NEOs away from the Earth, then it also has the ability to deflect them toward the planet.

Would that not be suicide, you might ask? Not necessarily. Imagine that nation A spots a hundred-yard asteroid that will just miss the Earth. Without telling anyone, it could divert that projectile, while still far away, in such a way that it will slam into enemy country B, wiping out its capital city with a hundred-megaton explosion. And nation A could claim no prior knowledge. Or country X might see another small asteroid heading for its own territory and, in self-defense, divert it in such a way that it hits nation Y, a whole ocean away.

This scenario is called the deflection dilemma. The ability to protect ourselves from impacts by NEOs is a double-edged sword. Because of this, many civilian scientists, including this author, argue that there is no need to build asteroid defense systems until an actual threat is identified. We should carry out an appropriate benign surveillance program first.

Forget Hollywood

Several movies have used as their story lines comets or asteroids hitting the Earth. Should they be taken seriously?

Hollywood rarely misses a good story. After the media fuss about asteroids and comets built up during the 1990s, several popular books were published that described the threat, including my own *Rogue Asteroids and Doomsday Comets*. A rash of TV documentaries followed. It was only natural that the big movie studios would try to dramatize the basic theme, leading to the blockbuster movies *Armageddon* and *Deep Impact*. But they had been preceded twenty years before by *Meteor*.

Meteor

The term meteor correctly applies to the streak one sees in the night sky when a tiny particle burns up on entering the atmosphere, but that is the least of the inaccuracies in this movie. Starring Sean Connery and Natalie Wood, *Meteor* is a mishmash of hocus-pocus science from start to finish, and unlike, say, *2001: A Space Odyssey*, it really shows its age. Predictably enough, the storyline involves a fictitious last-minute solution to an impending asteroid strike: blast it to pieces right above the atmosphere. Unfortunately, it just doesn't work like that.

Armageddon

Did the moviemakers do a better job two decades later? In terms of heightening public awareness of the issues, the answer is yes, and scientists concerned about the impact hazard, including myself, welcome that. But once again the science was way off.

Take *Armageddon*. We're told that an asteroid "the size of Texas" is heading for Earth. The problem is, we know of no asteroids that big.

Even Ceres, out in the main belt, is only about 600 miles (900 kilometers) across. The joke in the astronomical community at the time was that the script had originally said "an asteroid the size of Lubbock, Texas," but somewhere along the line the word "Lubbock" was dropped. The producers of *Armageddon* argued, perhaps with some justification, that the moviegoing public would simply not believe that a rock only a mile or so across could cause global devastation. But that's the way it is: that's a million-megaton bang.

Earlier, I explained how a comparatively small fraction of the sky is scanned monthly to the requisite faintness (the brightness required to discover asteroids through a specific telescope), but there we were discussing the search for half-mile asteroids. This Texas-sized asteroid would be a million times brighter. For a number

Above The asteroid impact theme is not new to Hollywood. In 1979, moviegoers were gripped by *Meteor*, and bought up associated merchandise.
Below Well-directed meteoroids batter Manhattan skyscrapers in *Armageddon*. In reality, cosmic projectiles this small would have been absorbed by the atmosphere tens of miles up.

ARMAGEDDON

of reasons, some of them obvious, it's stretching credulity to suggest such a beast could creep up on us unawares, as it did in the movie.

Leaving astronomy aside, the idea of being able to split such an object by boring to such a tiny depth into its surface is implausible. Scaling this behemoth down to the size of an orange, the storyline has the heroes drilling down a distance equivalent to penetrating the zest on the orange's exterior – not even the thickness of the pith comprising the rest of the skin. And the energy required to separate the two hemispheres is way beyond the yield of the nuclear bombs depicted in the movie.

Deep Impact

The science in *Deep Impact* is better, but the movie still has some serious shortcomings. A comet like the one depicted would not pass unnoticed. Comets are usually discovered, often by amateur astronomers, as soon as they are bright enough. It was no accident that Alan Hale and Thomas Bopp found their comet independently on the same night. Spotting comets is relatively easy. It's the much smaller and fainter asteroids that could sneak up on us unnoticed, with no warning until they actually hit.

Several other aspects of *Deep Impact* fudged the physics. One was the notion that the impact locations of the two cometary fragments could be predicted. Inert objects, such as asteroids, whose motion is governed solely by the gravitational pulls of the Sun and planets, are to an extent predictable. If we were provided with sufficiently precise orbital data, we might predict the impact point on the Earth for an asteroid to within a finite margin of error. But comets are different. The evaporation of their icy constituents produces a jet force that, when combined with the spinning, tumbling motion of the cometary nucleus, rapidly alters the comet's direction, making it virtually impossible to predict its precise path to better than some thousands of miles or kilometers. Not only would we not be able to identify the impact location for an active comet; we might not even know for sure whether it would hit or miss until a day or so before the calamity.

Movies like these make fine entertainment. But don't imagine for a moment that they are accurate dramatizations of what might happen. The reality is far, far more frightening.

Below Although it had several scientific faults, *Deep Impact* was closer to reality. Its depiction of the giant tsunami an oceanic impact would produce was phenomenal.

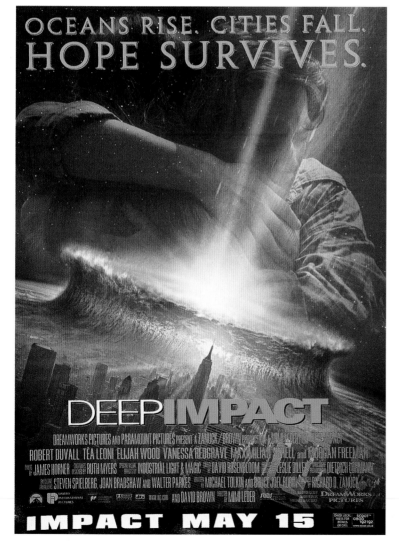

OCEANS RISE. CITIES FALL.
HOPE SURVIVES.

DEEP IMPACT

DREAMWORKS PICTURES AND PARAMOUNT PICTURES PRESENT A ZANUCK/BROWN PRODUCTION A MIMI LEDER FILM "DEEP IMPACT"
ROBERT DUVALL TÉA LEONI ELIJAH WOOD VANESSA REDGRAVE MAXIMILIAN SCHELL AND MORGAN FREEMAN
MUSIC BY JAMES HORNER COSTUMES DESIGNED BY RUTH MYERS SPECIAL VISUAL EFFECTS BY INDUSTRIAL LIGHT & MAGIC EDITED BY DAVID ROSENBLOOM FILM EDITOR LESLIE DILLEY PRODUCTION DESIGNER DIETRICH LOHMANN
EXECUTIVE PRODUCERS STEVEN SPIELBERG, JOAN BRADSHAW AND WALTER PARKES WRITTEN BY MICHAEL TOLKIN AND BRUCE JOEL RUBIN PRODUCED BY RICHARD D. ZANUCK
AND DAVID BROWN DIRECTED BY MIMI LEDER

IMPACT MAY 15

CHECK LOCAL PRESS FOR DETAILS OR CALL

The Spaceguard Foundation

● The NEO impact hazard affects the whole world, and demands an international solution.

Earlier we read about the political moves in the United States in the first half of the 1990s that led to an enhancement in that country's efforts to search for and track NEOs. Although the U.S. still dominates research in this area, the programs currently underway are too modest, both in the area of the sky they cover and the faintness they achieve, if we are truly serious about ensuring our safety.

Since the mid-1990s, the NEO impact hazard has made a number of appearances on the international agenda. But despite the encouragement of the United Nations that all member states should become involved, and the Council of Europe's 1996 motion calling for a pan-European program, little or no real action has resulted. There have, admittedly, been a few small initiatives here and there. In the face of the very real risks, however, this low level of activity is nothing short of scandalous.

A Problem for All Nations

No matter where on Earth it struck, an asteroid or comet of more than a certain size or energy would directly affect our entire planet, with dire consequences. These would include blast damage, firestorms, tsunamis, the poisoning of the atmosphere, and global cooling. The threshold size for such an outcome may be about a mile (1.6 kilometers), but it could be smaller. Even the 60-yard object that blew up over Siberia in 1908 seems to have cooled the northern hemisphere appreciably. That was chicken feed compared to what's out there.

This is not a national problem: it's a problem for all nations. We all live in, and depend on, the same global ecosystem, and we would all be subject to the gross environmental upsets a major impact would cause.

In 1996, seeing the need for an international solution, a group of astronomers established the Spaceguard Foundation, with its headquarters in Rome, Italy. We noted earlier that Italy is home to many of the best researchers in the field of NEO dynamics; the nation also has considerable

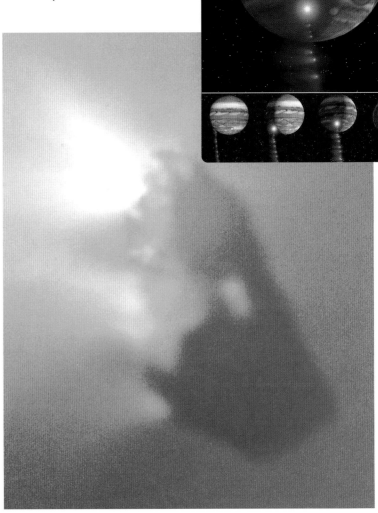

Above The nucleus of Comet Halley photographed by the Giotto spacecraft in 1986. Such an object hitting the Earth would cause a calamity of the same magnitude as that which killed the dinosaurs.

expertise in observing these objects. Given the need to spread responsibility outside of the United States, Rome seemed a good choice.

I was the Foundation's first vice president; today, membership includes most of the world's active researchers in the NEO field, no matter where they live. Many other individuals have joined simply because they feel it is a worthwhile cause. After all, we are trying to safeguard the future of the human race.

Local Branches

Apart from the central body in Rome, branches of the Spaceguard Foundation have been set up in other countries, including Germany, Croatia, Canada, and Japan. The names of these organizations vary locally due to legal differences, but basically they all adhere to the central aim of the foundation. That is, to ensure that the necessary search program is carried out such that, if there is an NEO due to hit us within the next few decades, we will find it far enough ahead of time for preventive action to be taken.

One of the most active national groups has been Spaceguard U.K., whose lobbying has been pivotal in ensuring that the impact hazard is kept in the British public eye. The U.K. government is now looking seriously at the threat posed by Near-Earth Objects. Perhaps the United States will soon have a true international partner in the quest for the killer asteroid.

Underside Uncovered

My own major contribution to the mission of Spaceguard was to establish and direct the NEO search-and-tracking program in Australia between 1990 and 1996. At that time, the Australian government terminated all funding for the project, and my team was disbanded. This was despite fervent lobbying by virtually the entire world community of scientists involved in Spaceguard, as well as other concerned parties.

This means not only that there is now no search program covering the southern sky, but also that NEOs discovered by the American search teams are liable to be lost due to the lack of backup tracking in the south. Australian residents are among the most at risk from NEO impacts through tsunami generation: about 90 percent of the population live in the coastal regions of the east, south, and west of that vast island nation, with those coasts facing out onto huge oceans. Most of Target Earth is Target Ocean.

Opposite The impacts by Comet Shoemaker-Levy 9 on Jupiter in 1994 should serve as a warning to us.
Below left To search out NEOs while they are still far from Earth, we need large telescopes to collect their dim light.
Below The crest of Spaceguard U.K., part of the international Spaceguard Foundation.

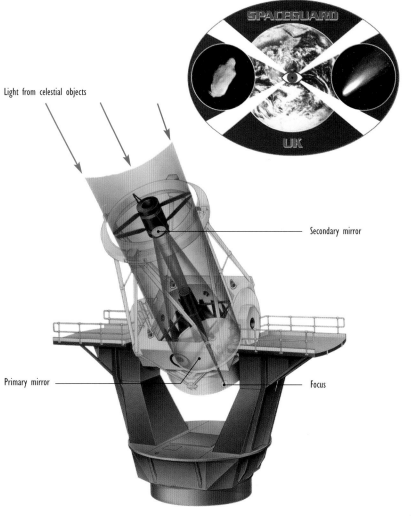

Light from celestial objects

Secondary mirror

Primary mirror

Focus

The Torino Scale

The Torino scale provides a mechanism for assessing the severity of the threat from any potential NEO impactor.

The decision to set up the Spaceguard Foundation was made at a conference, "Beginning the Spaceguard Survey," which was held on the island of Vulcano, just north of Sicily, in 1995. In 1999, we returned to Italy for a similar meeting in Turin.

At that conference there was discussion about how the media react to announcements of predicted possible impacts. For example, in early 1998 there had been some commotion when it was realized that an asteroid called 1997 XF11 might run into the Earth in 2028. Even with the initial data, the probability of a collision was deemed to be small, and it was soon shown to be close to zero: the asteroid will miss.

But how can one quantify such things? Which should we be more concerned about: a house-size rock that will certainly hit the Earth next week, or a 1-mile (1.6-kilometer) asteroid that has a one-in-100,000 chance of slamming into our planet in ten years' time? The former would kill few, if any, whereas the latter would claim a billion or more human victims. If it hit.

What Is Newsworthy?

There are so many variables that dealing rationally with each individual case becomes problematic. Say we found a 1-mile (1.6 kilometer) asteroid that we could prove will certainly hit the Earth in a century, should we be concerned about it or simply leave it to our great-grandchildren to deal with?

The thermometer of public opinion provides part of the answer to these questions, and this is itself affected by the selectivity of news editors. What they find suitable to print or carry in the electronic media shapes the public view. In 1992,

we thought for a short while that Comet Swift-Tuttle might come back to collide with the Earth in 2126 – 134 years into the future – and that hit the headlines all around the world. More recently several asteroids have been found to have small but finite probabilities of hitting us within the next forty or fifty years, but you read little or nothing about them.

After a while, yet another announcement about a potential impactor will lose its newsworthiness. It is possible to cry wolf too often. But if we don't alert people to the danger, then nothing will be done.

Scaling the Danger

Some astronomers involved in this field feel that they, the media, and the public in general need some form of scale that allows individual objects and the hazard they pose to be quantified. This

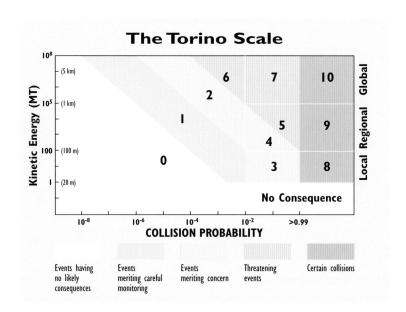

The Torino Scale

Assessing Asteroid and Comet Impact Hazard Predictions in the Twenty-First Century

Events having no likely consequences	**0**	The likelihood of a collision is zero, or well below the chance that a random object of the same size will strike the Earth within the next few decades. This designation also applies to any small object that, in the event of a collision, is unlikely to reach the Earth's surface intact.
Events meriting careful monitoring	**1**	The likelihood of collision is extremely low, about the same as a random object of the same size striking the Earth within the next few decades.
Events meriting concern	**2**	A somewhat close, but not unusual encounter. Collision is very unlikely.
	3	A close encounter, with 1 percent or greater chance of a collision capable of causing localized destruction.
	4	A close encounter, with 1 percent or greater chance of a collision capable of causing regional devastation.
Threatening events	**5**	A close encounter, with a significant threat of a collision capable of causing regional devastation.
	6	A close encounter, with a significant threat of a collision capable of causing a global catastrophe.
	7	A close encounter, with an extremely significant threat of a collision capable of causing a global catastrophe.
Certain collisions	**8**	A collision capable of causing localized destruction. Such events occur somewhere on Earth every 50 to 1,000 years.
	9	A collision capable of causing regional devastation. Such events occur between once per 1,000 years and once per 100,000 years.
	10	A collision capable of causing a global climatic catastrophe. Such events occur once per 100,000 years, or less often.

group includes Rick Binzel of the Massachusetts Institute of Technology. At the 1999 conference in Turin he proposed a scale that indicates – using numbers between 0 and 10 – the level of risk represented by a given NEO. This is now referred to as the Torino scale (see illustrations).

Some people have described it a "Richter scale for impacts," but that is misleading. The Richter scale is a direct measure of the energy released in an earthquake, and it is continuous. A jump of one on the Richter scale implies an increase in energy by a factor of ten.

The Torino scale, in contrast, is discontinuous. It has to be, because it quantifies several distinct parameters. First the probability of an impact is calculated using available data. Then the outcome of any impact is also factored in using the mass and speed of the NEO in question. For example a 5-mile asteroid could be classed only as a 0, 1, 2, 6, 7, or 10 on the Torino scale. If it did hit, it would cause global devastation, and in this case the different points on the scale simply give a sense of the likelihood of its hitting. Because the scale attempts to combine both a probabilistic calculation and a potential damage assessment, it is difficult to know whether you should be more concerned about a large object classed as a 2, or a much smaller one classed as a 3.

For this and other reasons, some scientists see the Torino scale as limited in usefulness. Only time will tell whether people actually want to use it.

Opposite and left The Torino scale provides a way to quantify the risk posed by individual asteroids and comets. Are we looking at a small body that will cause little damage or a large one certain to cause many deaths – but with only a small chance that it will strike the Earth? Impacts in Hollywood movies tend to focus on level 10 events: virtually certain collisions that will cause global disruption, and maybe a mass extinction.

8 Taking a Closer Look

If we are to protect ourselves from asteroids and comets, we must try to understand them better: to do this we need to examine them at close quarters using space probes.

If you could hold a chunk of a comet in your hand, what would it look like? Scientists cannot answer that question yet.

Take Comet Halley, for example. We have been observing it with the naked eye for more than two thousand years. In the 1830s astronomers sketched and studied it. In 1910 it was photographed for the first time. In 1986 thousands and thousands of photographs were taken, sophisticated investigations were made using ground-based telescopes, and a fleet of five spacecraft were sent to get up-close data. And yet we still don't know its mass and density to better than a factor of two or three!

Clearly this means that we cannot be sure how much damage such an object would do if it were to strike the Earth. We can only make estimates based upon certain assumptions. It is not too much of an exaggeration to say that, when it comes to diverting any NEO on a collision course with us, we don't really know whether we need an elephant gun or a butterfly net. We still know very little about their physical nature. But we can be sure that they have wreaked havoc in the past and will do so again in the future, unless we intervene. We must learn more about our foes.

Galileo, Incidentally

Five spacecraft went to spy on Comet Halley in 1986. Limited information has also been retrieved from two other comets. NASA's International Cometary Explorer probe flew by Comet Giacobini-Zinner the year before, but it had no camera on board. And after its Halley encounter, the European Space Agency's Giotto probe was sent on to Comet Grigg-Skjellerup in 1992. Alas, its camera had earlier been disabled. We still have a lot to learn about comets.

What about asteroids? On its way to Jupiter, the Galileo space probe flew past two main-belt asteroids – 951 Gaspra and 243 Ida – giving us our first up-close views of these objects. The images it returned contained few surprises: we were expecting elongated but smooth bodies, pockmarked by craters, and that's exactly what we saw.

It is not at all clear whether Earth-approaching asteroids are quite the same. NEOs are much smaller and will not necessarily have the same origin and history as main-belt asteroids. If we are to understand the threat from near-Earth objects then it is these bodies that we must study in detail.

Accessible Targets

There are many NEOs that would make attractive targets for a space probe. Quite apart from our understandable interest in their nature, they are the most accessible objects in space.

By "accessible" here I mean as regards the energy required to reach them. In space travel, this is by far the most important factor – the distance involved is a secondary consideration – and this energy is usually measured in terms of the change in velocity required. Or in the jargon of space travel, the "delta-vee."

The Moon might be the closest object for us to visit, but because of its significant gravity, a substantial delta-vee is necessary to land gently on its surface through the use of retro-rockets to slow descent, and even more to blast off again. An NEO is so small that it has no substantial gravitational pull, so that is not a consideration. The only delta-vee required is to match speeds with it, and there are several known asteroids with speeds relative to Earth of only 2 or 3 miles (3 or 5 kilometers) per second.

Quite apart from our interest in NEOs from both the scientific and the self-preservation aspects, they also represent sources of raw materials for our future exploitation of space. Forget mining the Moon; asteroids and comets contain all we need, and they are much easier to reach.

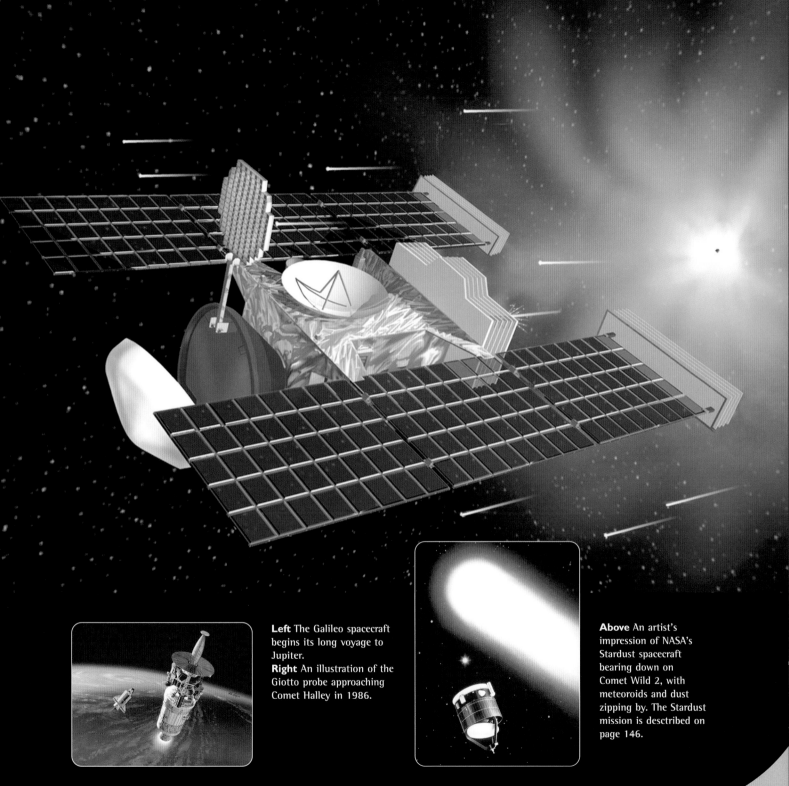

Left The Galileo spacecraft begins its long voyage to Jupiter.
Right An illustration of the Giotto probe approaching Comet Halley in 1986.

Above An artist's impression of NASA's Stardust spacecraft bearing down on Comet Wild 2, with meteoroids and dust zipping by. The Stardust mission is desctibed on page 146.

Near-Earth Asteroid Rendezvous

● The NEAR-Shoemaker space probe is the first dedicated asteroid mission.

The first near-Earth asteroid to be discovered was 433 Eros in 1898. It is appropriate that it is the target of the first specific spacecraft mission to an asteroid.

NASA's Near-Earth Asteroid Rendezvous, or NEAR, probe was launched in 1996, to take a circuitous path to Eros. In fact, the trajectory was somewhat longer than had been planned. NEAR was supposed to approach Eros slowly in early 1999, using a rocket burn to decelerate so it would rendezvous with the asteroid. In the event, a malfunction led to NEAR whizzing past Eros rather than slowing to meet it gently. The mission was saved because the two objects – the spacecraft and the asteroid – had orbits that would bring them together again in February 2000. A successful burn then inserted NEAR into orbit around Eros. By that time the craft had been rechristened NEAR-Shoemaker, in honor of the late Gene Shoemaker, who had long campaigned for space missions to asteroids.

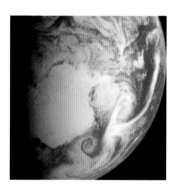

Above Two years after its launch, in early 1998, the NEAR spacecraft flew back by the Earth on its way to Eros and took this image of the Antarctic surrounded by swirling vortices of cloud.

Waltzing Mathilde

Along the way, in 1997, NEAR had taken advantage of the proximity of its path to a main-belt asteroid, 253 Mathilde, to gather data on that celestial object. After Gaspra and Ida, Mathilde was the third asteroid for which we obtained up-close images. These turned up some surprising results. It was larger than we had thought: because it is so dark, previous determinations of its size, made from its brightness when viewed through a telescope, turned out to be underestimates. Mathilde is actually larger than Gaspra or Ida. But the big surprise was its density, which was very low.

We tend to assume that asteroids are solid rock, and yet there is evidence from their spin rates that most are actually rubble piles. Imagine a heap of boulders in space. If the heap is rotating, the centrifugal force will be trying to make the boulders fly away from one another. In the absence of anything to "stick" them together (for example, ice between the boulders which might act like glue), there is only self-gravity to oppose the centrifugal force. It happens that, for a spherical rubble pile as hypothesized here, the self-gravity and centrifugal forces are equal when the period of rotation is about two and a half hours. Any faster and a rubble pile will fly apart. Almost all asteroids for which we have measured rotation rates – by seeing how their brightnesses vary – have spin periods of more than five hours. That makes us think that they are loose clusters of separate components, their apparently smooth surfaces produced by smaller rocks and dust.

When the NEAR spacecraft flew past asteroid Mathilde, the gravitational pull of the target allowed us to derive its mass and density. That turned out to be only about 60 percent of the density of stony meteorites or common rocks on Earth. This implies that there must be voids within the asteroid, and adds weight to the view that many asteroids could be rubble piles.

Above A true-color composite image of Eros shows how it would look to an astronaut's eyes. The asteroid is about 20 miles (30 kilometers) across.
Opposite above A montage of images of Eros obtained as the NEAR-Shoemaker probe homed in on the asteroid in early 2000.
Opposite center The relative sizes of the two asteroids visited by NEAR: Mathilde on the left, Eros on the right.
Opposite below An infra-red image of Eros tells researchers about its surface composition.
Below A huge crater on Eros. The energy of the impact that formed this concavity must have been almost enough to shatter the whole asteroid.

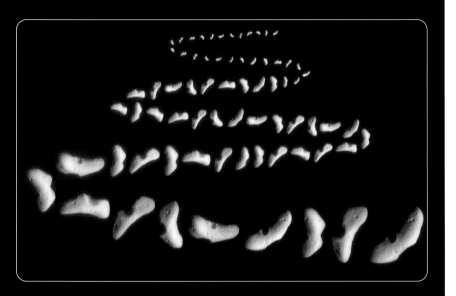

Encounter with Eros

The spacecraft flybys of Gaspra, Ida, and Mathilde returned only a handful of detailed images, and with only a limited range of geometries. NEAR, in orbit around Eros, has been able to map the entire surface of the asteroid in detail, from many different altitudes and sun angles. It is planned eventually to bring the spacecraft down to within 30 or 40 yards or meters of the surface, which will give unprecedented resolution.

We have already made many unexpected discoveries, including craters resulting from impacts large enough to have shattered the entire object into fragments if it were a rubble pile. So maybe Eros is a monolith.

But apart from the familiar dents in its surface caused by meteoroid impacts, Eros also has protrusions. Small mounds that look like large boulders have been spotted here and there, and their origin is a mystery. They could be large meteoroids that have landed gently on Eros, held there by its weak gravitational field.

Maybe they are part of the underlying rock, sticking out through a dusty coat. As the NEAR-Shoemaker mapping of Eros continues, we hope the answer will be revealed.

Above An artist's impression of the NEAR-Shoemaker space probe.
Top The launch of the NEAR mission from Cape Canaveral in February 1996.

Advanced Technologies

New space probes are already on their way to asteroids and comets, and more are being prepared for launch in the next few years.

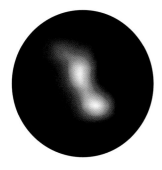

Above Although Deep Space 1 (DS1) was successful from the perspective of technology testing, the scientific return from its encounter with near-Earth asteroid Braille was fairly sparse. This is the best image obtained of the target.
Far right A poster showing the two types of target that DS1 was planned to visit: an asteroid (Braille) and a comet (Borrelly).
Right This spoof of a magazine cover highlights the new technologies tested on the DS1 mission.

We are entering the golden age of asteroid and comet exploration. Over the next decade, led by NEAR-Shoemaker, spacecraft operated by the U.S., European, and Japanese space agencies will visit a dozen NEOs. As we shall see in the next few pages, there is even a commercially operated asteroid mission due for launch soon.

Many of these missions are made feasible by the utilization of new, advanced technologies, developed through such programs as Star Wars. Autonomous systems allow satellites to make their own navigation decisions, lightweight instruments minimize the launch costs, and microelectronics allow more processing power to be packed on board.

Deep Space 1

NASA's Deep Space 1 (DS1) mission is a prime example of the application of new technology. The program was designed with technology testing in mind, rather than scientific exploration. This was the first deep space mission to employ an ion drive engine for propulsion. These drives had been used previously, to a limited extent, in Earth-orbiting satellites, but never for an interplanetary voyage.

In general, rocket motors are fired for a limited period, during which they generate a large thrust, substantially changing the spacecraft's speed. An ion engine achieves the same effect in quite a different way. It spits out

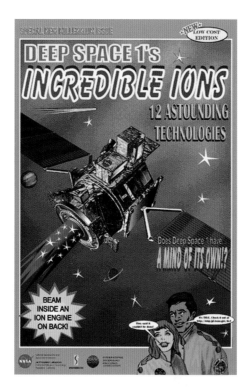

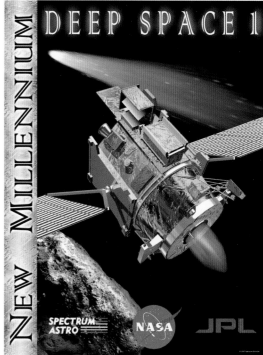

a stream of high-speed ions producing a very small thrust, but this weak thrust is maintained for a long period of time – months or years, compared to the seconds or minutes of a conventional rocket burn. Overall the ion engine generates more thrust per unit mass of fuel, in part because it uses relatively heavy atoms like xenon as the propellant and accelerates these to a high speed in a magnetic field before allowing their escape through the rear-facing nozzle. This makes these engines a more economic proposition, as long as you are not in a hurry. Since a voyage to an asteroid may take a year or so, speed is not a problem.

DS1 successfully used an ion engine to fly past near-Earth asteroid 9969 Braille in July 1999. Although the pictures it returned were disappointing, it showed that the system works. The schedule for DS1 is not finished yet, however. It is now en route to an encounter with periodic Comet Borrelly in September 2001. After nearly a century of observing this comet by telescope since it was discovered in 1904, we can now look forward to a much closer view.

Deep Space 4

Several other missions in the Deep Space series are planned. Deep Space 4 (DS4) was slated to visit a comet and analyze a sample, but that project is currently on hold.

The original DS4 was merged with a mainly European craft called Champollion to form what became known as Space Technology 4. Champollion was a comet sampler originally intended to be part of the European Space Agency's Rosetta mission, which we will discuss shortly. It was named after Jean-François Champollion, a French scholar who deciphered the Rosetta Stone in 1822, making it possible for us to translate the language of ancient Egypt.

The idea was that the Champollion module

would land on the surface and drill through the thin, inert crust of the comet to investigate the volatile material thought to lie just below. We know that comets are largely composed of organic chemicals. It may be that comets were the source of these molecules on early Earth, making the formation of life possible. Clearly there is a need to know just what comets contain within their solid cores.

Perhaps this mission will be reinstated soon. Another NASA mission that will expose the subcrustal layers of a comet, although in a quite different way, is Deep Impact. That's a comet probe we'll discuss in a few pages.

Above An artist's impression of the Champollion module landing on the surface of a comet to take a sample. Champollion is a mainly European component of NASA's Space Technology 4 mission, which is currently on hold pending funding.

Bounding Around an Asteroid

● **Both Japan and a commercial U.S. venture plan missions to near-Earth asteroids.**

Above The shoebox-sized nanorover, Muses-CN, designed to leap like a grasshopper around the surface of an asteroid.

Some NEO discoveries provoke special interest among researchers, not because the asteroids pose a danger of impact, but rather because their particular orbits around the Sun mean that they are relatively easy targets for space missions.

For example, if an NEO has an orbital plane quite close to that of the Earth, less energy is required to send a probe to rendezvous with it than would be the case if its orbit were highly inclined. Similarly, if the NEO has a near-circular orbit, this also reduces the energy requirement.

Considerations like these have led us to realize that some NEOs represent the most accessible objects in space, easier to get to than the Moon. For our future exploitation of space, asteroids and comets may be the most economic sources of raw materials. And that makes them valuable commodities.

Prospecting for Space Minerals

A growing recognition of the economic value of NEOs is part of the motivation behind a space mission called NEAP (Near-Earth Asteroid Prospector), scheduled for launch in 2002. The target asteroid has not yet been chosen.

NEAP is a commercial venture under the control of the SpaceDev, Inc. of California. On board will be various scientific instruments operated and paid for by several U.S. universities. The satellite will also carry a compact disc containing the names and greetings of various subscribers.

This mission is the brainchild of Jim Benson, the principal of SpaceDev. He realized early on that, as well as being a potential threat to us, NEOs could be a boon in terms of the natural resources they contain. Benson intends to lay claim to a near-Earth asteroid by landing on it. The legality of this claim has still to be tested, as several international treaties apply to outer space. But as the old saying goes, possession is nine-tenths of the law.

Japan Has a Plan

Another spacecraft being readied for launch to a near-Earth asteroid is Muses-C, built and operated by the Institute of Space and Astronautical Science

Below and bottom The Near-Earth Asteroid Prospector (NEAP) spacecraft, planned to fly to its target in 2002 and stake a claim.

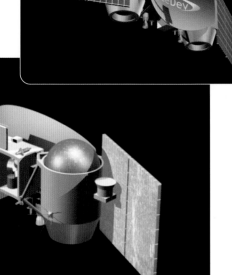

in Japan. The designated target asteroid is presently known as 1998 SF36.

This asteroid, about a half mile (800 meters) in size, was discovered in September 1998 by the LINEAR project team (see page 110). It frequently passes close by the Earth, with its next approaches due in March 2001 and June 2004. This will allow astronomers to scrutinize the asteroid with telescopes before the spacecraft reaches it not long after.

Muses-C is due for launch in November 2002, arriving at its target three years later. On board the mothership will be two small landers that will examine the asteroid's surface close-up. One is a Japanese device named Minerva, while NASA has built a tiny rover currently called Muses-CN; one would imagine that a more prosaic name will be found for it soon.

The plan is that this rover – more often called a nanorover by dint of its small size – will navigate around the asteroid surface not so much by turning its wheels, but more by jumping like a grasshopper. The asteroid has such weak gravity that an astronaut would be able to jump right off it. By applying a precise amount of upward acceleration, the nanorover should be able to bound around the surface in leaps of a hundred yards or meters at a time. The idea is simple: the nanorover jumps up, and during the several minutes before it lands again, the asteroid will have spun beneath it. Sounds like fun!

Above An impression of the Japanese Muses-C craft dropping its NASA-built nanorover onto the surface of an asteroid.

Stardust

● **NASA's Stardust probe should tell us more about the origin of the very stuff of which we are made.**

Above Comet Wild 2, the target of the Stardust probe. An artist's impression of Stardust approaching the comet is shown on page 139.

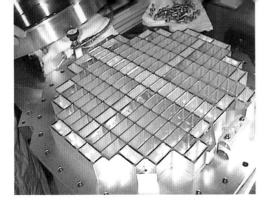

Above The pallet of separate aerogel cells flown on Stardust to collect cometary dust grains intact.

After the universe formed, the first stars contained only hydrogen, with a little helium synthesized through nuclear fusion reactions in the stellar cores. Bigger stars could produce other atoms, but none heavier than iron. The heavier elements, like lead and uranium, were made in phenomenal supernova explosions. These spit out into interstellar space huge clouds of gas and dust from which new stars form.

The Stuff of Life

The recycling of stellar material made available the raw materials from which the Sun, the planets, moons, comets, and asteroids were made. And us. The notion that we are stardust is to a large extent true.

The Earth had a hot, torrid beginning. Formed from a vast number of large lumps of rock and ice coming together in enormously powerful collisions, it was depleted of all its volatile material, like water and organic molecules, which would have rapidly vaporized. These compounds must have been reintroduced to our planet later, making the evolution of life possible. But how? We think the answer is through comets, which seem to be largely made of these chemicals.

To confirm the hypothesis, we need more information about cometary compositions. We need to bring samples of a comet back to our laboratories and work out their chemistry and origin. That is the aim of the Stardust mission.

Through the Maelstrom

To land on a comet, grab a chunk of it, and bring it back to Earth for analysis presents a huge

Right The Stardust probe is readied for launch. Gold foil is used to wrap most of the satellite to aid thermal stability and also to shield it to some extent from meteoroid impacts. The technician wears protective clothes to prevent contamination of the spacecraft.

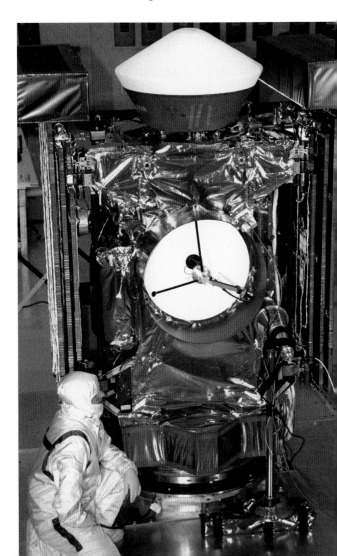

technological challenge. To land on a comet you have to match speeds with it; that would require a phenomenal amount of fuel and pose many other problems. The ices sublimating from a solid cometary nucleus would produce an extremely daunting environment in which to work; and remember, there wouldn't be an astronaut on board to take control. The spacecraft would be many minutes of radio travel time away from Earth, so remote-control is not an option: it would have to be a fully autonomous robot.

A more feasible mission would entail flying by a comet with a probe able to collect some of the cometary dust that surrounds it. This in itself is no piece of cake, akin to flying through a maelstrom of gas, dust, and meteoroids, but in terms of complexity it is less challenging.

You might even say it would be easy, as long as one major problem could be overcome. How can you collect dust grains intact when they are hurtling toward you at 10 miles (16 kilometers) per second? That speed gives them more than enough energy to vaporize totally on impact. To save them from instantaneous destruction you need some device that will slow them down gradually.

The answer to this quandary is the wonder-material called aerogel. Aerogel is the closest you can have to nothing while you still have something. It has an extremely low density, meaning that tiny grains are gradually decelerated from their high impact speeds, and it is a phenomenal thermal insulator.

The Stardust probe, launched in February 1999, has a container full of separate aerogel cells. Some of these have been exposed in flight to collect samples of interplanetary dust. By orienting the spacecraft in the appropriate way relative to its cruise path, another exposure period of a few months may collect some interstellar dust (particles from outside the solar system). Dust captured during this phase cannot

Right Aerogel is a wonderful insulator. Here crayons on an aerogel slab escape melting, despite the heat from the blowtorch applied to its underside.

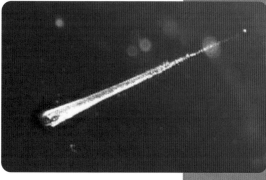

Above A single track made by a dust particle fired into aerogel. The sample remained intact, decelerated gradually by this unique substance.

be in permanent orbit around the Sun because it would not be moving fast enough to catch up with the Stardust probe from behind. If successful, this will deliver our first ever samples of material from outside the solar system.

The main aerogel cells will be uncovered only during the phase of the spacecraft's encounter with its target comet, called Wild 2, in January 2004. After that they will be closed up again for the two-year-long cruise home. If all goes well, the capsule containing tiny fragments of that comet will gently come to Earth, aided by a parachute, in the western United States in January 2006. Then the chemists can get to work.

Above Tracks made by tiny particles of dust shot into aerogel at very high speeds in a laboratory test.

STARDUST

147

Rosetta

● **Rosetta is a European mission designed to follow a comet to see how its activity changes as it approaches the Sun.**

Above The Rosetta Stone in the British Museum, from which the Rosetta spacecraft derives its name.

We read earlier about the Rosetta Stone and the Frenchman Champollion who first deciphered the hieroglyphics carved on it by comparing them with the Greek translation also inscribed on the stone. This gave us the key to understanding the pictograms that festoon Egyptian buildings, providing much of our knowledge of that era. In much the same way, the Rosetta mission will enable astronomers to unlock many of the secrets of the origin and evolution of the solar system, maybe even of life itself.

The original plan was that Rosetta would carry the Champollion module, bringing a sample of the nucleus of a comet back to Earth for analysis. This was deemed too costly and extravagant, and the mission will now be somewhat simpler and more modest in scope. But it will still address a very important question: how does a comet behave when subjected to increasing solar heat as it approaches the Sun?

Turning the Water On

Already a decade and a half in the making, Rosetta is an ambitious spacecraft. Over that period the destination comet has been altered several times. The current – and, we hope, final – target is Comet Wirtanen, which was discovered in 1948 from the Lick Observatory in California.

Comets appear bright in the sky because they are surrounded by large clouds of gas, most of which starts out as frozen water in the comet nucleus, although other icy constituents such as ammonia, carbon dioxide, and many organic chemicals also evaporate as the Sun heats them. Water is the dominant constituent, though, and its physical properties are such that it starts to sublimate when the comet is about 3 AU from the Sun. This is why comets often become suddenly brighter as they pass this threshold distance.

Rosetta's mission is to rendezvous with a comet when it is still beyond that critical distance and, while flying alongside it, gather data on how the nucleus springs into life as solar radiation makes water and other icy materials turn to gas. Rosetta will get to Comet Wirtanen in late 2011, when the comet is about 3.5 AU from the Sun.

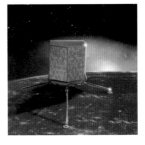

Top A design model of the Rosetta lander, named Roland.
Above An artist's impression of Roland on the surface of Comet Wirtanen.
Below An Ariane rocket will be used to launch Rosetta, from the European Space Agency's base in French Guiana.

TAKING A CLOSER LOOK

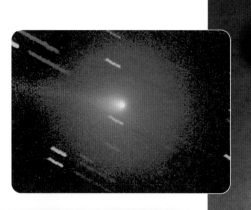

A Lifetime's Work

Not only has Rosetta been a long time in the planning; its execution will also be a lengthy affair since the mission is not due to conclude for another thirteen years, with further data analysis to follow. In fact, those involved in the program from start to finish will have spent a whole career working on it. More than most people, they will be hoping to avoid last-minute hiccups.

Launch is scheduled for January 2003. The first encounter with a celestial object will be in May 2005, when Rosetta will use the gravity of Mars to accelerate it back to a flyby of the Earth in October, its speed increased through the slingshot effect. It will do precisely the same thing two years later, in October 2007, having passed by asteroid 4979 Otawara in the meantime.

After the second Earth encounter, it will have enough energy to reach Comet Wirtanen at the end of November 2011, with at least one asteroid flyby (of 140 Siwa) planned during that four-year cruise phase.

If all goes according to plan, Rosetta will fly alongside its target for more than eighteen months as Comet Wirtanen comes to perihelion in July 2013, all the while monitoring how the comet's activity alters. It will drop an instrument package onto the surface to make direct measurements of the composition and structure. If successful, the lander will inspect several locations by hopping from one place to the next.

Top left Comet Wirtanen, the principal target of the Rosetta probe.
Center left Rosetta meets Comet Wirtanen in 2011 and photographs its surface before the Roland lander begins its close-up investigations.
Above Rosetta in flight, with its solar cells extended. The background is an image of Comet Halley obtained by the Giotto probe in 1986.

Right The European Space Agency will install a new radio dish near Perth in Western Australia to receive data from Rosetta.

The Diversity of Comets

NASA plans to direct two further spacecraft to comets in the next few years.

The Stardust spacecraft currently winging its way to reach Comet Wild 2 in 2004 is the first NASA space probe dedicated to exploring a comet. But we know that comets come in all shapes and sizes. There is a diversity in comet behavior that we must understand.

If you needed to save your life from an attacking lion or tiger, having studied your neighbor's pet cat might help, although perhaps not a lot, despite their all being classified as felines. Similarly, if we want to defend ourselves against a comet, we need to know more about the spectrum of objects covered by that generic term. We've made a start. Two new spacecraft planned by NASA will take us a lot further.

Deep Impact

From the Earth, our view of a comet's nucleus is mostly obscured by the coma: the cloud of vapor that surrounds the solid lump in the middle. Even if we push a spacecraft in close, like Giotto or Rosetta, we will still see only the crust, the material stable enough to avoid evaporating under solar heating.

What is under that dark surface? Answering that fundamental question is the aim of the Deep Impact probe. Stealing its name from the motion picture, its mission is to create a deep impact – on a comet.

One way to discover what lies under the crust would be to land a robotic probe, have it grip the surface to hold on tight – remember that comets have very low self-gravity – and drill down through the upper layers. Technologically and economically, this is a horrendous task. An alternative would be simply to blow a hole through the crust. No explosives are necessary. A big lump of inert metal traveling at 5 or 10 miles or kilometers per second will suffice.

And that's exactly what is proposed. Deep Impact will release a half-ton cylinder of copper as it approaches Comet Tempel 1 in July 2005, with the expectation that it will blow out a crater at least 120 yards or meters wide and maybe 25 yards or meters deep. That will uncover the pristine material beneath, which is expected to sublimate as the sunlight penetrates the hole. All this will be monitored by the mother ship from a safe distance, as well as by Hubble, and by a host of astronomers using

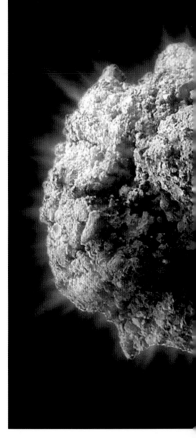

Left This artist's impression of the Deep Impact spacecraft shows the mother ship observing Comet Tempel 1 as it is hit by the half-ton (500-kilogram) projectile launched shortly beforehand.
Right A model of the Deep Impact spacecraft. The gold-covered cylinder at bottom is the projectile it will fire at the nucleus of the comet.

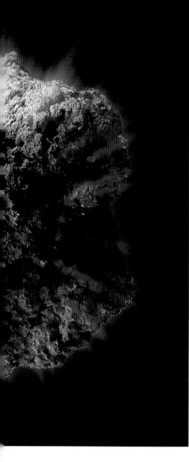

Earth-based telescopes. I need hardly add that this is scheduled to occur on July 4th. Expect some celestial fireworks.

A Comet Nucleus Tour

Contour, the acronym for the Comet Nucleus Tour spacecraft, is scheduled to visit at least three comets during the next decade. Two of them are among the most famous comets in our inventory: Encke and d'Arrest.

Although comets nowadays are named for their discoverers, we've seen that Edmond Halley was not the first to spot the comet bearing his name (1P/Halley – the "P" means "periodic"). He simply studied its motion and predicted its return. Similarly, the second comet in the master list, 2P/Encke, was not discovered by Johann Encke, the German astronomer who worked on its dynamics in the 1820s, but by the French astronomer Pierre Mechain in 1786. The French-German linkage was reversed in the case of 6P/d'Arrest, since Heinrich d'Arrest, who discovered it from Leipzig, was German despite his French name.

After its launch in July 2002, Contour is scheduled to approach within about 60 miles (100 kilometers) of 2P/Encke in November 2003. Perhaps it will render vital clues about how this object managed to get into such a small orbit, with its trajectory bringing it back near the Earth every forty months. The small orbit of Comet Encke is shown in the diagram on page 14.

Contour's next target also has a German name: 73P/Schwassman-Wachmann 3 (or simply SW3) was the third periodic comet discovered by this pair of astronomers, from Hamburg in 1930. SW3 was recently observed breaking into several fragments, giving us hope that when the spacecraft gets there in mid-2006 it will be possible to distinguish between old and relatively young surfaces on the major portion.

Above A science-based model for the appearance of the nucleus of Comet Tempel 1, the target of the Deep Impact space probe. This nucleus is thought to be about a mile or kilometer across, but no one knows for sure.

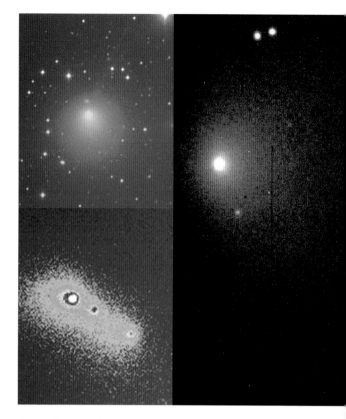

Above The three target comets for the Contour mission. At top left is Comet Encke, which has the most frequent returns of any known comet; at bottom left is Comet Schwassmann-Wachmann 3, which was recently observed to break up into multiple components, as in this image; and on the right is Comet d'Arrest, clearly showing the cloud of dust that surrounds it.

Another two years will pass before the final comet on the Contour itinerary, 6P/d'Arrest, is reached in August 2008. That will bring an end to the nominal mission. Of course, at the rate at which asteroids and comets are now being discovered, it is not at all unlikely that some other potential target will be found in the meantime, extending the tour.

The Nemesis Rock

An NEO is found on a collision course with Earth. How could we intervene?

Most likely there is no large asteroid or comet due to hit us soon. Like most animals – and most humans – in the Earth's history, we will go through our lives without suffering the severe consequences of a major impact.

But what if our number comes up? What if, to our misfortune, an NEO – let's call it the Nemesis Rock – is on its way to wreak havoc within the next decade or two?

First, the bad news: at our current level of effort in looking for these lethal projectiles, the chances are that we would not see it prior to impact. Unless you count eight or ten seconds' warning as constituting an adequate time to prepare. We would probably have no inkling of our fate until it was too late.

But let us again imagine that the urgings of bodies such as the Spaceguard Foundation succeed in waking up the world's governments, persuading them to make the investment that will ensure we are not caught off guard. What would we do then when the Nemesis Rock is spotted and recognized for what it is, perhaps with ten years' advance warning?

Making Sure

First, we would need to make sure we knew what we were dealing with. When you are staring down a loaded gun barrel, there's nothing you won't do to save your life. Similarly, no expense is too great to avoid asteroidal Armageddon. We would launch not

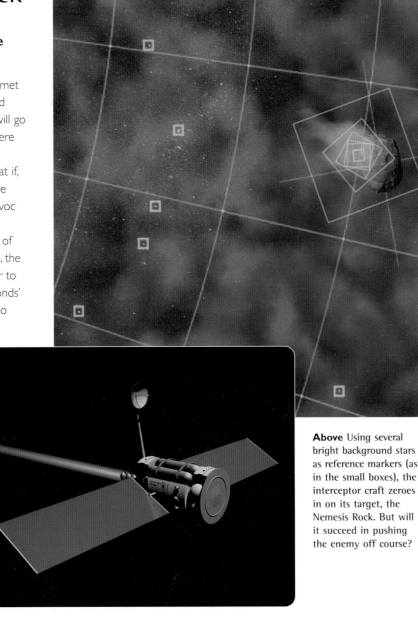

Above Using several bright background stars as reference markers (as in the small boxes), the interceptor craft zeroes in on its target, the Nemesis Rock. But will it succeed in pushing the enemy off course?

Above A hypothetical interceptor craft is sent to tackle the Nemesis Rock. Using an ion drive powered by solar cells, it is able to deliver a multimegaton nuclear weapon to the dangerous asteroid.

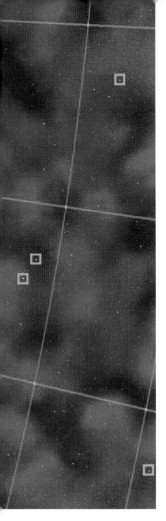

just one, but a whole fleet of space probes to investigate the nature of the beast. The first might carry a radio beacon to drop onto its surface. That way we would be able to determine our enemy's orbit with better precision, confirming its Earth-intercept path and the optimum direction in which to give it a shove. Next, other probes would take a closer look and investigate the composition and structure of the Nemesis Rock. Is it solid stone? Or metal? Or a rubble pile?

Perhaps it's a comet in disguise? That is, might it be a comet that has formed an insulating crust, so that there is ice below the rocky, tarry surface? We have seen comets switch off, such as Comet Wilson-Harrington, which was active (that is, evaporating volatile materials) in the late 1940s, but had apparently become inert and asteroidal by the time it was accidentally rediscovered in 1979. On the other hand, Comet Encke must have been dormant – looking like an asteroid – in the many centuries leading up to its discovery in 1786, because otherwise it would have been found earlier. No telescope would have been necessary to spot it if it was as bright as it has been in the two centuries since.

We should hope that the Nemesis Rock is a dormant comet. If so, simply smashing through its crust with a Deep Impact-type projectile would probably be enough to divert it to miss

the Earth. The escaping gases would provide sufficient thrust to shift the reawakened comet off its collision course.

The Last Resort

But it is more likely that the approaching impactor would be an inert lump of rock. As noted in our earlier example, if you contract a deadly disease, none of the possible remedies are pleasant. To divert an asteroid heading for Earth would almost certainly require the use of nuclear weapons.

By igniting a suitable device above the asteroid's surface, a layer perhaps a yard or meter thick would be evaporated off the face turned toward the explosion. The jet force would be enough to maneuver the asteroid to miss the Earth, provided the intercept were performed far enough ahead of time.

That, at least, is the theory. The physics equations all work, but putting it into action is another matter. Should we do a practice run, or do we just rely on our good fortune and technological prowess?

I'll leave you to ponder that question. I'm sure of one thing. To have a fighting chance of avoiding an impact catastrophe set up by the clockwork of the heavens, we have to find our enemy quickly. There is no excuse for not carrying out the necessary space surveillance program. We must have an answer to that vital question:

Above Could we divert the Nemesis Rock using a powerful nuclear device?

Is there a big one due to hit us soon?

world determine to prevent such a disaster happening again. Although the chances are statistically low, the consequences are just too high to gamble with the future of civilization.

On the eve of 2001 – the real dawn of the new millennium (what *has* all the fuss been about?) – many commentators have filled their column inches with prophecies of doom and gloom. Some feel that science and technology are moving ahead too fast. Others say we've not come far enough: for instance, manned probes are not yet on their way to Saturn, as my late friend Stanley Kubrick and I envisioned in *2001: A Space Odyssey*.

Well, we were too optimistic for 2001 – but perhaps too pessimistic as to when Spaceguard would begin. It is already happening long before 2077.

Not so long after *Rendezvous with Rama* appeared, geologist Walter Alvarez was investigating the rock strata laid down near Gubbio in Italy 65 million years ago – around the same time the dinosaurs became extinct. He found high concentrations of iridium, which is common in meteorites but rare on Earth. With his father Luis (a Nobel Prize winning physicist and one of my oldest American friends – see dedication of *Glide*

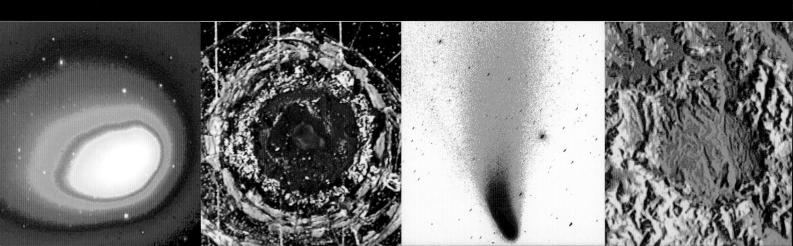

Path), Walter and his team developed a theory that the dinosaurs – along with many other species – died as the result of a huge asteroid impact.

There is an interesting coincidence here. In *Rama*, I had written about an asteroid wiping out Padua and Verona, and sinking Venice forever – and then Walter, digging just next door, found evidence of an actual historical impact!

Such cosmic catastrophes are both fascinating and terrifying, making them compelling subjects for novels, TV documentaries, and movies. I myself revisited the theme in *The Hammer of God* (1993) – which Steven Spielberg optioned a couple of years later, before making *Deep Impact*.

But that's fiction. In *Target Earth*, Duncan Steel tells the actual story of how our planet has been hit many times by asteroids and comets, often with results far worse than Tunguska. More importantly, he shows how it will be devastated again — unless we act now. That's what Project Spaceguard is all about.

Convincing the governments of the world and the public on this hazard has been a long and difficult task. I've done my bit by being a patron of the Spaceguard Foundation and its various allied organizations. But it requires all sane people to pressure their leaders into making Spaceguard a reality. We need to look to survival strategies for 3001, not just 2001.

Four power-packed words in the English language end with the letters "dous": they are tremendous, stupendous, horrendous, and hazardous – all of which describe the effects of asteroid and comet impacts on Target Earth.

To adapt a slogan from the environmental movement: it's not every day that you have an opportunity to save a planet.

Sir Arthur C. Clarke
Colombo, Sri Lanka
August 2000

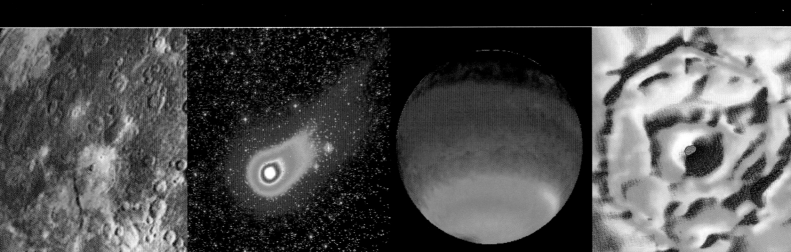

Glossary

Achondrites Stony meteorites not containing chondrules; they are therefore largely homogeneous.

Albedo The fraction of impinging sunlight reflected by a celestial body such as a planet, moon, or asteroid.

Asteroid An inert rocky or metallic body in orbit about the Sun, ranging down in size to roughly 10 yards or meters. Synonym: *minor planet.*

Astronomical Unit The mean distance of the Earth from the Sun, about 93 million miles or 150 million kilometers. Abbreviation: *AU.*

Basin In the context of impact cratering, a very large bowl-shaped crater on a planet or moon that has later been largely filled with other rock.

Black Hole A celestial object with such a strong gravitational field that light cannot escape from it.

Caldera The depression formed by the collapse of the center of a volcano.

Chondrites Stony meteorites containing chondrules.

Chondrules Small round masses, mainly made of the minerals olivine and pyroxine, thought to be some of the earliest material to solidify as the solar system condensed from the pre-solar nebula.

Coma The vast cloud of gas and dust produced around the solid nucleus of a comet when sunlight causes the more volatile constituents to sublimate.

Comet An object in orbit about the Sun composed not only of rock, but also of volatile materials such as ice and organic chemicals that may sublimate when subjected to heating by sunlight. This produces the characteristic coma and tails that a comet displays.

Corona and chromosphere Two of the upper layers of the atmosphere of the Sun, most easily seen during a total solar eclipse.

Doppler effect The shift in frequency of sound waves or electromagnetic radiation resulting from the relative motion of the source and the observer (e.g. the change in pitch of an ambulance siren heard as the vehicle passes by).

Eccentricity A measure of the deviation from circularity of an orbit. The planets have low-eccentricity orbits around the Sun, whereas many comets and asteroids have high-eccentricity and, therefore, elongated orbits.

Ejecta Rocks and other material ejected from a crater in the impact process.

Fireball A bright meteor or shooting star.

Galaxy A very large group of stars, together with clouds of gas and dust, bound together by mutual gravity. Our galaxy, the Milky Way, contains more than 100 billion stars, of which the Sun is one.

Impactite Glassy residue formed from target rock and meteoritical material melted in an impact event.

Kinetic energy The energy possessed by a body due to its motion.

Lagrangian points Locations in space where the gravitational pulls from different objects (e.g. the Sun and Jupiter) cancel each other out, providing some stability.

Main-belt asteroid An asteroid orbiting the Sun as a member of the major concentration of such objects between Mars and Jupiter.

Mare (plural: maria) One of the large dark plains on the Moon. These are impact basins later filled with basalt from volcanic eruptions.

Metamorphism Changes in the structure of rock caused by heating or pressure. Impacts cause shock metamorphism, altering the target rocks in characteristic ways.

Meteor A fiery streak in the sky produced by the arrival of a meteoroid in the upper atmosphere. Synonym: *shooting star.*

Meteoroid A rocky body in space less than about 10 yards or meters in size and, therefore, too small to be described as an asteroid. Meteoroids smaller than about a thousandth of an inch (a few tens of microns) are termed *interplanetary dust.*

Near-Earth Asteroid An asteroid with an orbit that brings it within about 0.3 AU of the terrestrial orbit around the Sun. Abbreviation: *NEA.*

Near-Earth Object Similar to Near-Earth Asteroid, but includes comets. Abbreviation: *NEO.*

Nebula A large cloud of gas in space; may also contain dust.

Nucleus The solid core of a comet, composed largely of ice but also rock and volatile materials such as organic chemicals, plus solid carbon dioxide and ammonia.

Oort cloud A spherical distribution of billions of comets that extends to a quarter of the distance to the nearest stars (about 100,000 AU). The source of many observed comets.

Paleontology The study of fossil animals and plants.

Perihelion The point in an object's orbit where it comes closest to the Sun. (The furthest point is termed *aphelion.*)

Photosphere The luminous visible layer of the Sun.

Planetesimals The small bodies from which it is believed the planets agglomerated. Many asteroids, especially those in the main belt, are left over planetesimals that never became incorporated into any of the planets.

Prograde Orbital motion in the same direction as the planets: counterclockwise when viewed from the north.

Protoplanet A planet that is forming; similar meaning to planetesimal.

Retrograde Orbital motion in the opposite direction to the planets: clockwise when viewed from the north.

Satellite An object in orbit about the Sun or a planet. May be used to refer to either a natural body, such as a moon of one of the giant planets, or a man-made spacecraft.

Sedimentation The settling and compaction of material to form a type of rock (e.g. mud on the bottom of the sea).

Solar system The group of celestial objects tied to the Sun by its gravity, including the Earth, the other planets, asteroids, and comets.

Tektite Small glassy bodies believed to have been formed from melted terrestrial rocks that solidified into aerodynamic shapes as they flew through the atmosphere after a major impact.

Tidal locking The situation in which an object maintains the same orientation as it orbits (e.g. the Moon keeping the same face directed towards the Earth at all times).

Trans-Neptunian Object One of the several hundred large asteroids/cometary bodies found since 1992 beyond the orbit of Neptune. Collectively known as the Edgeworth-Kuiper belt.

Tsunami A massive oceanic wave produced by an undersea landslip, earthquake, or perhaps an NEO impact. Often erroneously called a tidal wave.

Index

Page numbers in *italics* refer to illustrations

Credits

Quarto would like to thank and acknowledge the following for supplying pictures reproduced in this book:

Key: l left, r right, c center, t top, b bottom

Courtesy John Africano, AMOS/Haleakala Observatory: p108tr & br; p109br Copyright Anglo-Australian Telescope Observatory/Photo: David F. Malin: p73bl; p78tl (insert) Courtesy Armagh Observatory: p94tr & cr The Art Archive: p12br; p148tl David Asher: p6t; p7t Courtesy Ball Aerospace, Boulder, Colorado: p150bl Courtesy Jim Benson, SpaceDev, Inc.: p144cr & br Courtesy Richard Binzel, Massachussetts Institute of Technology: p136; p137 J.P. Bradley, MVA Associates: p73t J.C. Casado: p75t Catalina Sky Survey, University of Arizona: p112cl J-L Charmet: p151 & tr Corbis: p101 (main pic.) David Crawford, Sandia National Laboratories, New Mexico: p64tl (main pic.), tr (insert); p122t & c; p127b Don Davis: p23tc; p31b; p47tr; p49l; p54tr (insert); p62; Dominion Astrophysical Observatory: p112tr Courtesy Dutch Meteor Society: p74 (main pic. and insert) Sara Eichmiller-Ruck: p79b European Southern Observatory: p12tr; p89t; p102cr (insert) Courtesy of the European Space Agency: p148tr, cr & br; p149 (all pics.); p134cr; p139br; p148tr, cr & br; p149 (all pics.) Courtesy Peter S. Fiske, Lawrence Livermore National Laboratory, California: p85tr Alan Fitzsimmons, Queens University, Belfast: p25cr & br; p101bl & bc Gordon Garradd (Loomberah Observatory): p13tl; p78tr (insert); p107 James Garry, Fastlight: p145 Courtesy Tom Gehrels: p104tl Courtesy Geological Survey of Western Australia: p95bc & br Courtesy of Victor Gollancz, an imprint of Orion/Ill.: Peter Mennim: p155tr Giles Graham, Open University: p80br; p81tl (insert) William K. Hartmann: p26tl; p29t; p51cl (insert), p64b; p124 (main pic.); p125tl David Harvey, University of Arizona: p151tr Mark Herbert, University of Kent: p73br; p81bl, bc & br Alan Hildebrand, University of Calgary: p63cl & bl; p83tr; p98br Courtesy of Hodder Headline: p96c Image Bank: p65 (insert); p68b Courtesy The Isaac Newton Group of Telescopes, La Palma: p89bc & br Gary Kitley, Klet Observatory: p106br McDonald Observatory, University of Texas: p40c Scott Manley, Armagh Observatory: p9t; p14; p17; p18; p21; p22; p25tl; p27tl; p114c & br; p115tr; p129; p152t & b; p153 Courtesy Alain Maury: p104tr; p113br Courtesy Karen Meech, University of Hawaii: p146tl National Aeronautics and Space Administration: p9br (insert); p10bl; p11; p16br (P.Thomas & B Zellner, Space Telescope Science Institute); p19tr; p20cl; p24tl (R. Albrecht, ESA/ESO Space Telescope European Coordinating Facility): p26cl (Photo: H. Weaver & P. Feldman), b (Photo: H. Weaver & T.E. Smith, Space Telescope Science Institute); p30bl & tr; p31tl; p32cr; p33t, c & b; p34tr, l (JPL/Northwestern University; p35tr (JPL/Northwestern University); p36t, c; p37b; p38t & b; p39l & r; p40tr (Space Telescope Science Institute); p41c, t (Space Telescope Institute);

p42bl (LASCO consortium of the SOHO satellite/ESA), tr (LASCO consortium of the SOHO satellite/ESA/Montage by Steele Hill); p43t & b (LASCO consortium of the SOHO satellite/ESA); p44tl, tr & br; p45tl, cl & tr; p46t, c & b; p47tl & bl; p48t, b (P.Thomas & B. Zellner, Space Telescope Science Institute); p49tr; p51 (main pic.); p52r (main pic.); p55l & r; p59; p60bl & br; p61tr, c & bl; p63t (Painting: Don Davis); p71; p80tr, bc & br; p81 (main pic.); p92tr (Painting: Don Davis); p96b; p99l, tr (Painting: Don Davis); p102l (main pic.), tr (Space Telescope Science Institute); p108bl (courtesy Eleanor Helin, Jet Propulsion Laboratory); p109t (courtesy Eleanor Helin, Jet Propulsion Laboratory); p116c; p125bc & cr; p127t; p128; p134t; p139t & bl; p140l, tr & br; p141 (all pics.); p142 t, c & r; p143; p144tl, p146tr&br; p147 (all pics.); p150tr & br National Astronomy and Ionosphere Center, Cornell University/National Science Foundation: p117tr National Radio Astronomy Observatory: p116br National Research Council of Canada: p112tr Courtesy Richard O. Norton: p82tl, tr & br; p83cl, cr & br Courtesy of Orbit Books, an imprint of Little, Brown and Company (UK)/Ill.: Fred Gambino: p155cr Courtesy Steve Ostro, Nasa/Jet Propulsion Laboratory & Scott Hudson, Washington State University: p20cl; p117; p118l; p119l, c & r; p120l & r; p121 (all pics.); p131tr Courtesy Richard & Roland Pelisson, SaharaMet, France: p82cl; p83bl; p85cl Pictor: p29b; p65 (main pic) Courtesy Royal Greenwich Observatory: p82tl, tr & bc; p83cl, cr & br Jim Scotti, Spacewatch, University of Arizona: p112bl; p151tl Martin Setvak, Klet Observatory: p106c Yan On Sheung: p75b Siding Spring Observatory, Australian National University: p40bl Sky & Telescope magazine: p114tl Courtesy Space Guard UK: p135cr Courtesy Spacewatch, University of Arizona: p9bl; p104cl; p105 (main pic. & insert) Courtesy Pavel Spurny: p78 (main pic.) Courtesy Grant Stokes, Lincoln Laboratories: p110; p111tl, tr & br Milos Tichy & Jana Ticha, Klet Observatory: p20bl; p96tr; p102 (main pic.) Huraj Toth, Comenius U. Bratislava, Modra Observatory: p76 Touchstone Pictures/The Ronald Grant Archive: p132b Courtesy Chris Trayner, University of Leeds: p87t F. Tsikalas, S.T. Gudlaugsson, J.I. Faleide & O. Eldholm, University of Oslo: p57tr & cr UIP/The Ronald Grant Archive: p133 University of California, Berkeley: p32tr U.S. Department of the Interior: p68tl Courtesy of U.S. Geological Survey: p53tr (insert); p63cr & br; p67br; p103br; p130bl U.S. Navy: p130c; p130tl Courtesy Wendee Wallach-Levy: p103tr Steven N. Ward, Institute of Tectonics, University of California at Santa Cruz: p69r Courtesy Alexander Zaitsev: p117cr.

Quarto would also like to thank the Natural History Museum, London, for providing the tektites photographed on page 67.

All other photographs and illustrations are the copyright of Quarto Publishing plc. While every effort has been made to credit contributors, Quarto would like to apologize should there have been any omissions or errors.